GUIDELINES FOR
Failure
Investigation

Developed by the
Task Committee on Guidelines for
Failure Investigation
of the Technical Council
on Forensic Engineering

Howard F. Greenspan, Fellow, ASCE, Chairman 1987-89
James A. O'Kon, Member, ASCE, Chairman 1984-87
Kimball J. Beasley, Member, ASCE
Joseph S. Ward, Fellow, ASCE

Principal Author, James A. O'Kon

ASCE
1852

Published by the
American Society of Civil Engineers
345 East 47th Street
New York, New York 10017-2398

ABSTRACT

The *Guidelines for Failure Investigation* by the Task Committee on Guidelines for Failure Investigations of the Technical Council on Forensic Engineering, American Society of Civil Engineers provides an overview to the functions and responsibilities of the successful failure investigation. The *Guidelines* covers all aspects of the investigative process from the planning stage, the development of a failure hypothesis, and the preparation of a concise report, to the engineer's role as an expert witness. To help explain this process, two investigations in specific engineering disciplines, geotechical and structural, are developed in detail. With this complete coverage, the *Guidelines* not only provides an introduction to the investigative process for the initiate forensic engineer, but also supplies the veteran investigator with references and guidance.

Library of Congress Cataloging-in-Publication Data

Technical Council on Forensic Engineering (American Society of Civil Engineers). Task Committee on Guidelines for Failure Investigation.

    Guidelines for failure investigation/developed by the Task Committee on Guidelines for Failure Investigation of the Technical Council on Forensic Engineering.

    p. cm.

    Includes bibliographical references.

    ISBN 0-87262-736-5

    1. Structural failures—Investigation. 2. Forensic engineering. I. Title.

T4656.T38   1989

624.1'7—dc20   89-39537

                                             CIP

TABLE OF CONTENTS

i. INTRODUCTION

Much credit belongs to the design and investigative engineers who have made contributions to a built environment, thus elevating the American standard of living to the highest on earth. The dramatic structures of tall buildings, the dynamic construction of dams, and the complex infrastructure of cities have shaped our incredible nation scape. Everything in our built environment is expected to operate as designed; but when things do not function, such as the failure of a dam, a power blackout, or even a leaking roof, we are reminded that things do not operate continually or forever.

A new discipline has been created to deal with the investigation of failures and performance problems in the built environment. This discipline requires of the engineer the full spectrum of scientific skills as well as exemplary qualities of character. This new discipline is known as Forensic Engineering.

To assist in the enhancement of the body of knowledge possessed by this rising profession, the American Society of Civil Engineers commissioned their Technical Council on Forensic Engineering to develop a guide. To accomplish that task, the Technical Council on Forensic Engineering formed the Task Committee on Guidelines for Failure Investigations; this guide is the result. It is devoted to the functions and responsibilities of the forensic engineer, as well as methodologies for conducting a successful failure investigation.

The charge of the Task Committee on Guidelines for Failure Investigation was to review the existing publications relating to failure investigations; for example, "Guide to Investigations of Structural Failures", by Jack R. Janney (ASCE 1979), and "Masonry Failure Investigations", by Clayford T. Grimm, MASCE (ASTM 1982). Our review of these two documents, as well as other references, found that they are valuable tools for identifying and preparing an investigation of a failed or a distressed structure. However, they do not prepare the forensic engineer for other critical elements of an investigative commission, such as the development of a comprehensive report or, possibly the most difficult part of the forensic process,

~~the preparation for and participation in legal~~
proceedings.

This Guide, therefore, is intended to augment and
expand on existing publications by providing a
complete outline of the forensic engineer's
responsibilities (i.e., presenting guidelines for
general investigative techniques, and outlining
the methodology for writing reports and preparing
for and participating in the legal process). Its
aim is to aid the forensic engineer in the pursuit
of his or her professional responsibilities.

ii. PURPOSE

The purpose of these guidelines is to provide an overview of the qualifications and responsibilities that are required of a forensic engineer; to provide recommendations for logical methodologies that can be used to identify causes of failure or performance problems in constructed facilities; and to describe the role of the forensic engineer as an expert witness during the legal process. These guidelines are also intended to provide an introduction to the investigative process for the initiate forensic engineer, as well as to provide references and guidance for veteran investigators.

The methodology and investigative techniques contained in this guideline will lead the investigator through the complete course of a forensic commission, starting with the actual incident of failure. It will traverse the investigative process, from the preparation of an investigation plan, the development of a failure hypothesis, and the preparation of a concise report, to the engineer's role as an expert witness in the legal process. The guidelines are not limited to investigation of catastrophic failures or distressed structures that endanger life; they address a range of failures, including serviceability problems, which are often as costly in dollars as catastrophic failures.

The long range goal of these guidelines is to assist all civil engineering disciplines in the process of failure investigation by using an interdisciplinary methodology; thus, professionals acting as a team can optimize their collective talents and ascertain the cause of a failure in any aspect of the built environment. The process for investigation presented in this guide is general. However, investigation of failures related specifically to the structural and geotechnical engineering disciplines is developed in detail. Later editions of this guide will include background data relating to the failure investigations of other engineering disciplines, including materials of construction, transportation and environmental investigations.

We wish to thank Kenneth Carper of ASCE's TCFE for his excellent editing of this work, Clayford T. Grimm for his work relating to the investigation of masonry failures, attorney Phillip Fortune for his assistance in preparing the section on the legal process, and Elizabeth Greenspan for her editorial review of the manuscript. But we especially wish to express our particular gratitude to James A. O'Kon who produced the major portion of the manuscript.

Howard F. Greenspan
Chairman, Task Committee
on Guidelines for
Failure Investigation

CHAPTER 1

FUNCTIONS AND RESPONSIBILITIES OF THE FORENSIC ENGINEER

The practice of Forensic Engineering may be defined as the application of the engineering sciences to the investigation of failures or other performance problems. While some commissions undertaken by the forensic engineer may not involve sworn testimony in a legal forum, many do. When commissioned to serve as an expert witness the engineer must possess special knowledge in order to serve as an expert before the courts of law, or in arbitration or administrative proceedings.

The forensic engineer translates knowledge gained through engineering education and experience into an opinion that he or she presents to a judge and jury of laymen. It is they who make the ultimate decisions on matters of dispute.

Forensic engineering includes investigation of the physical causes of failures, as well as other sources of claims and litigation. This includes, but is not limited to, the following: preparation of engineering investigation reports, testimony at hearings and trials in administrative or judicial proceedings, and the rendition of advisory opinions to assist in the resolution of disputes affecting life or property.

The qualified professional who desires to enter the ranks of forensic engineering should be fully aware of the responsibilities and stressful situations encountered by the forensic engineer when serving as an expert witness during a legal contest; the opinions of the forensic engineer, the results of the failure investigation and the engineer's qualifications and capabilities will be under judicial review during the litigation process. Exposure to the stresses of a legal contest may be found distasteful by some individuals.

To prepare the forensic engineer for the legal process, Chapter 6.0 discusses in detail the role of the forensic engineer when serving as an expert witness in a legal dispute. Also provided is an overview of the legal process and the activities related to the services of the forensic engineer.

1.1 QUALIFICATIONS OF THE FORENSIC ENGINEER

There are two prime qualities that must be attributed to a forensic engineer. This individual must be an expert in his or her field and impartial as to cause and responsibility during the course of the investigation, and responsible for testimony in judicial proceedings.

As an expert, the forensic engineer must have a thorough knowledge of the subject under investigation. This knowledge must have been acquired not only by formal education but by years of practice; the forensic engineer must be able to relate the present situation to past ones in order to determine the cause. As an expert, the forensic engineer uses all of his or her technical resources in the finding of facts.

The forensic engineer must always avoid assignments that differ from his or her expertise. As to impartiality, the forensic engineer must avoid the appearance of conflict of interest, bias, or advocacy. The forensic engineer arrives at final conclusions on the basis of sound engineering fundamentals and from the evidence that has been developed during the investigation. Forensic engineers must display no bias during the course of their assignments and always conduct themselves as professionals in expert witness testimony.

The practice of forensic engineering involves the full spectrum of services, ranging from investigation of the failure to participation in the legal proceedings that often follow. Therefore, the qualified forensic expert must be a licensed, experienced engineer thoroughly familiar with the nature and type of engineered facility being investigated, including design, materials, construction techniques, and operations of the facility. The expert also should be familiar with building codes, test methods, contractural arrangements and the economics of construction. In addition, the forensic engineer must have completed the required level of education, should participate in professional societies, and publish works. Personality characteristics such as objectivity, respect for confidentiality, and integrity are fundamental.

The qualifications and capabilities required of the forensic investigator should be presented in the engineer's resume or curriculum vitae. Development and preparation of a curriculum vitae are discussed in Section 1.2.

1.1.1   KNOWLEDGE OF THE FIELD OF EXPERTISE

The forensic engineer should pursue professional engineering activities that will lead to his or her acceptance as an acknowledged expert in the chosen field.

This expertise may be based on a background of experience in design, construction, investigations, education, professional activities and published works, as well as an extensive term of practice in the area of specialty. In addition to a particular area of expertise, a good investigator should have an understanding of the interdisciplinary engineering systems that interface with his or her specialty. The expert must strive to attain capabilities in reasoning and analysis that go beyond building codes, specifications and the simplistic models of engineering behavior used for design, in order to develop an intuitive understanding of systems and how they actually behave and why they fail.

1.1.2   EDUCATIONAL BACKGROUND

Because the credibility of the forensic engineering experts and the opinions expressed by them directly relates to their educational background, experts should be educated in the specialty that they practice. The experts should make every effort to attend short courses and seminars in their specialty to ensure that they are conversant with the latest technology and developments. Attendance at seminars should be noted in the resume or curriculum vitae. Experience as a visiting lecturer and advisor to universities, or as a speaker to professional groups in the field of specialization, contributes to the engineer's credibility as an expert.

1.1.3 LICENSE AS A PROFESSIONAL ENGINEER

A professional engineering license is essential to the credibility of the forensic engineer, and mandatory if the investigation results in the preparation of remedial measures for the distressed facility.

The credibility of the expert during an investigation and the weight given to testimony as an expert witness will be enhanced if the engineer is licensed in the state where the incident of failure occurred. While it is not absolutely necessary, it is to the advantage of the forensic engineer to be registered in the various states where a potential commission might be offered.

To facilitate registration in many states, it is recommended that the engineer obtain a NCEE certificate of qualification.

1.1.4 PUBLISHED WORKS

Acceptance as an expert in a specific field is enhanced if the professional has published works in his or her field of specialization. The variety of formats is extensive; it includes technical journals of professional societies, industry related magazines, textbooks and professional manuals. The subject of a published article need not be academic; it can be practical in nature.

1.1.5 MEMBERSHIP IN PROFESSIONAL SOCIETIES

Membership and participation in professional societies will establish the expert as a leader and activist in his or her field. Professional societies support research and disseminate technical information. Active participation in a professional society will indicate that the expert is well versed in state-of-the-art concepts through attendance at meetings, conferences, conventions, and has reviewed journals and other publications issued by the professional society.

Participation on a professional society's working committee indicates peer acceptance and suggests that the expert is a leader in his or her field. Membership in the American Society of Civil Engineering (ASCE), the National Society of Professional Engineers (NSPE), and other professional societies is important. In addition, the forensic engineer should attain membership in societies or committees that recognize their members as accomplished forensic experts, such as the National Academy of Forensic Engineers (NAFE), the ASCE Technical Council on Forensic Engineering, or forensic engineering committees established by other professional societies.

1.1.6   **CONFLICT OF INTEREST**

To maintain credibility in the investigation as well as to prevent impeachment in court, it is essential to avoid any encounter that may be considered a conflict of interest. Elimination of conflicts of interest will defuse accusations of favoritism during the investigation and suspicion during the legal dispute. Potential conflicts of interest could include the following:

1.1.6.1   **Association with the Project**

It is important that the expert witness not be previously associated with the project that is being investigated.

1.1.6.2   **Elimination of Bias**

The expert should evaluate and assess each commission to ascertain if previous associations with aspects of it suggest a conflict of interest. These past associations with the project could include the client, its

litigants, or the client's
attorneys. For example, bias
could be claimed if the expert
witness had been previously
retained by the litigant or if
the expert had a special relationship with the attorneys,
even though there may, in
fact, be no bias.

1.1.7 QUALITIES OF CHARACTER

To maintain all aspects of credibility,
the forensic engineer must demonstrate
qualities of character that will ensure
impartiality and avoid impeachment.
These qualities include objectivity,
confidentiality, honesty and integrity.

1.1.7.1 Objectivity

The forensic engineer must
maintain objectivity during
the investigation of a
failure. Objectivity starts
with the acceptance of an
assignment and extends through
the advice and opinions
provided to his or her client
during the preparation for
litigation. The expert must
strive to maintain integrity
and cognizance in order to
conduct an impartial investigation.

In some cases the expert is
retained by a client with
special interests. Whether
done consciously or not, it is
possible that the client's
interests could affect the
expert's thought process and
opinions, and the expert
should consciously strive to
remain objective. If objectivity is not possible, then
the commission should be ter-
minated.

The forensic engineer should
provide advice to the client

without prejudice or bias. As
the findings of the investigation are uncovered, the
forensic engineer should
apprise the client of the
actual situation in order to
provide the client with
sufficient time and opportunity to assess liability,
devise solutions, and plan
contingencies. The key to
ethical conduct as a forensic
engineer is one overriding
professional dictum: The
substance of opinions conveyed
by an expert should be the
same no matter who has
retained the expert.

In accepting a forensic
engineering engagement, it
must be determined what will
be expected from the client.
Will the expert be permitted
to be objective and develop
information, data and
materials that are necessary
to uncover causes of the
failure and convey the truth
as it is perceived? Will the
expert be required by the
client to support the client's
position no matter what the
investigation reveals? The
difference between these two
positions is the difference
between honesty and perjury.

The experts should reject any
assignment that does not fall
in their area of expertise,
that presents a conflict of
interest, or requires their
objectivity to be compromised
by the client's special
interests.

1.1.7.2   Confidentiality

The old expression, "loose
lips sink ships," is directly
applicable to a failure in-

vestigation. The findings or
results of the investigation
must be kept confidential
until the expert is directed
otherwise by the client or the
client's attorney.

The forensic engineer must not
discuss any culpability with
regard to the failure outside
the confines of the client and
the investigation team.
Special care must be exercised
when dealing with repre-
sentatives of the news media.
In all events, the forensic
engineer should rely on the
advice of legal counsel
regarding any discussions of
findings of the investigation
team prior to the conclusion
of litigation.

1.1.7.3 ## Honesty and Integrity

The forensic engineer must
have unwavering honesty and
integrity at all times
throughout the investigation,
as well as the ability to
remain composed under hostile
examinations.

1.1.7.4 ## Communication Skills

It is important for the expert
to develop the capacity to
speak and write convincingly
in clear, understandable
language. Written and oral
statements issued by the
expert should use common
language that eliminates
unnecessary technical jargon.

The most important reason for
using understandable language
is that the findings or
testimony of the expert must
be understood by the client or
a jury. They will not have a
technical background and will

understand the issues only
when they are discussed in
understandable language.
Chapter 3.0 discusses
techniques for clear
communication.

1.2 QUALIFYING THE EXPERT: THE RESUME

The qualifications of the forensic expert should
be expressed in a curriculum vitae or resume. The
curriculum vitae is a document that summarizes the
education, background and experience of a
professional. Because the curriculum vitae plays
a major role in establishing credibility in a
specific field, it should have a clear and concise
format, be well organized and be as complete but
as brief as possible. Its organization should
reflect the approach and philosophy of the
forensic expert.

The curriculum vitae should delineate the
engineer's design projects as well as forensic
engineering projects. When listing forensic
assignments, commissions for the plaintiff's
clients should be approximately balanced with
commissions for defendants. The object of this
balance is to provide the reader with an overview
that indicates a background with a broad base of
commissions and clients. This will help reduce
criticism from the opposing side in a legal
contest that has an engineer appearing as an
expert witness; the balance of projects and
clients will be a positive statement, countering
potential charges by the opposing attorney that
the engineer specializes as an expert witness, and
is appearing in court as a "hired gun" without the
benefit of a broad base of engineering design
expertise.

1.2.1   **Photographs**: Include a photograph of
         yourself in a conservative suit.

1.2.2   **Identification**:   Name, title and
         business address.

1.2.3   **Profession**: Title within organization
         and the area of specialization (e.g.,
         President, John Doe and Associates,
         Consulting Structural Engineers).

1.2.4 **Professional Registration**: List states where registered as well as registration numbers.

1.2.5 **Education**: List degrees and honors and the dates of awards; it is not necessary to list college activities or college fraternities. The exception would be collegiate honors and membership in honorary societies.

1.2.6 **Lecturer**: List universities, conferences, and seminars or panels where lectures were presented and appearances made. The full titles of the subject matter may be included.

1.2.7 **Published Works**: List in chronological order all published works; format this data as in a bibliography, stating title, date, publisher, and publication.

1.2.8 **Professional Society Membership**: List and describe all memberships in professional societies and delineate all active participation (e.g., officer, chairperson, or member) of committees within each organization. List all past offices held in professional societies.

1.2.9 **Awards**: List all awards relating to your profession. In addition, list recognition that has resulted from leadership activities in the profession, e.g., "Who's Who in America" listings, design awards, professional society awards, etc.

1.2.10 **Civic and Service Activities**: Provide listings of membership and leadership roles in civic and political organizations.

1.2.11 **Experience**: Provide a description of experience in your specialty in order to provide an overview of your management, design, and forensic background. Then list your employment record with duties and types of assignments carried out under each engagement.

1.2.12 **Representative Projects:** To provide an overview of your design and investigative experience, provide a list of about ten projects that illustrates a wide range of expertise. Include projects that may be recognizable to the reader. In addition, combine the projects to indicate a balance between forensic assignments and design assignments.

1.3 RESPONSIBILITIES OF THE FORENSIC ENGINEER

To provide an overview of the total responsibilities required for the proper execution of forensic engineering commissions, this section will outline the salient tasks that could be encountered during the course of an investigation (Figure 1).

The scope of responsibilities on a given commission may terminate at the completion of the investigative report; or the scope of services may extend beyond the initial report and include recommendations for remedial measures and/or preparation for a legal contest and participation as an expert witness in the civil litigation process.

This section describes in detail the types of clients that can be encountered; outlines recommendations for contracts for professional services; discusses development and management of the interdisciplinary team; and provides a summary of the responsibilities typical of the investigation, and typical of preparation for the role as an expert witness during formal litigation or dispute resolution.

1.3.1 CLIENT INTERFACE

The forensic engineer must carefully consider the client and the client's requirements prior to executing a contractual agreement.

1.3.1.1 Type of Clients

When a failure occurs the parties affected by the failure will vary in their interests, as well as in their

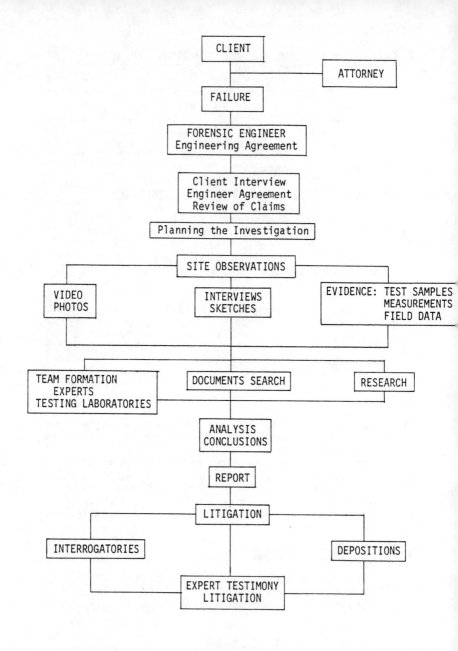

CRTICIAL PATH OF RESPONSIBILITIES OF FORENSIC ENGINEER

FIGURE ONE

-16-

number. The individuals or groups who may contract for the services of a forensic engineer might include the following:

1. Governmental

   Federal, state, and local authorities often retain forensic engineers for commissions related to failures of publicly owned facilities, or for engineering studies related to public interests. In addition, they may require consultants for condemnation proceedings, or for engineering investigations concerning the potential replacement of a failed or degraded facility.

2. Owner/Developers

   This category could include developers, corporations, financial institutions, educational institutions, real estate companies or other types of owners in the private sector.

3. Users

   This includes organizations or individuals who are leasing property or equipment and have experienced failures or performance problems.

4. Insurers

   The insurance companies and their agents who insure property, insure individuals for liability, or insure for bonding purposes might be the clients. They may require technical consultation in preparation of a claim, or in defense of a claim. They may also commission an investigation to identify and pursue actions against the party responsible for causing a failure.

5. Architects and Engineers

This category includes commissions by design professionals who may contract directly with a forensic engineer to carry out an investigation prior to the involvement of an insurance company, or to obtain the opinions of a forensic engineer serving as a third party in a design dispute.

6. Courts

Courts in certain jurisdictions will retain experts for evaluation of construction problems or failures and may also designate a forensic engineer as a court appointed expert to review the findings of experts for the litigants.

7. Contractors

This group includes material suppliers and manufacturers of equipment as well as general contractors and subcontractors. These entities may retain forensic engineers for investigation of technical and procedural errors leading to failures, material testing and the investigation of cost and construction delay disputes.

8. Attorneys

Forensic engineers might be retained directly by attorneys who are representing any of the groups listed above, or by an individual or a group of private citizens. The attorney may also serve in the role of an indirect client working on behalf of a litigant. In some cases, the attorney will be retained by an insurance company with the understanding that the insurer

pays all of the costs. Working
directly for the attorney usually
allows the consultant's work
product to be sheltered from
discovery (see also Chapter 6).

9. Public

Public groups or individual
members of the public may retain
forensic engineers for the
investigation of building failures, performance problems with
buildings or equipment, and
personal injury cases. These
public clients may retain an
expert to investigate and testify
at public hearings.

1.3.1.2 Initial Contact with the Client

The initial conference with the
client should be held to discuss
definition of the problem, preliminary identification of the scope of
work, and to determine whether or not
the assignment falls within the area
of expertise of the forensic engineer. In addition, this meeting
should be used to uncover potential
conflicts of interest, and to clarify
an agreement for services and compensation.

The consultant will request a briefing of the issues to be investigated
and the type of expertise that is
required by the client. The consultant will also clarify the
position that a commission cannot be
accepted if the client requires an
adversary role. If the consultant
has never served as an expert in the
specialized field under investigation, let the client know at that
time.

During the initial contact, a number
of issues can be discussed and
resolved in order to determine
whether or not it is in the best
interests of either party to spend

-19-

additional time discussing the matter. Determine if there are potential conflicts of interest because of past service with the attorneys for the opposition, the project in general, or relationships with any of the principals involved in the dispute. Determine if the commission involves a recent failure or if the initial problem occurred some time in the past and has advanced to a dispute; if litigation has begun, request a briefing of the status of litigation. If the case is near the trial stage, ask if an expert had previously been retained, and what became of the expert. Request the expert's name and make contact in order to ascertain why he or she is no longer involved.

The above information will enable the forensic engineer to ascertain if the client or the nature of the service requested will satisfy the standards established by the profession and permit the forensic engineer to accept the commission.

1.3.1.3 **Advice to the Client**

In addition to the role as failure investigator, the forensic engineer should discuss his or her value to the client in the role of technical advisor. As advisor, the forensic engineer may help the client understand the technical issues of the investigation, assist in evaluating its strong and weak points, and advise in the preparation of complaints.

1.3.1.4 **Compensation: Forensic Engineering Agreement**

The client should know in advance the approximate cost of each phase of the commission and of the total commission. The fee schedule, approximate budget, and nature of the

compensation should be identified at
the initial visit.

Some experts charge an hourly fee or
per diem basis for services, with
special rates imposed for on-call
service, and higher rates for
services during deposition testimony
and in court. The expert may be paid
on an hourly or daily basis, or as a
lump sum fee. Contingency fees are
unethical for engineers to accept,
and should not be considered; the
outcome of the case should have no
bearing on the expert's compensation.

To avoid misunderstanding, it is
usually necessary to execute a
forensic engineering agreement. In
any case, do not pursue an engagement
unless you have a valid contract
assuring appropriate compensation to
be paid in a timely manner, clar-
ification of your services, and protection of your work product.

1.3.2 THE INVESTIGATION

The planning and implementation of a
successful investigation involves a series
of interrelated tasks that are carried out
by the forensic engineer and the interdisciplinary investigation team. The
successful implementation of an investigation begins with the development of a
logical investigative plan and continues
with the selection of an investigation team,
site visits, development of testing programs, document and literature search,
synthesis of data, and development of
failure hypotheses. The investigation
usually culminates with the preparation of a
comprehensive report detailing the results
of the investigation.

The general methodology for conducting a
successful investigation is described in
Chapter 2.0. Chapter 4.0 describes the
methodology for conducting a geotechnical
failure investigation, and Chapter 5.0 describes the process for conducting a structural failure investigation.

1.3.3 REPORT PREPARATION

The comprehensive report of an investigation will provide a narrative and graphic overview of the investigation including the history of the project and the incident of failure, design considerations, analysis of the mode and causes of the failure, hypotheses on the causes of the failure and, in some cases, recommendations for remedial measures. A description of the various types of reports and a methodology for preparation of reports is contained in Chapter 3.0.

1.3.4 LEGAL CONSIDERATIONS

The responsibilities of the forensic engineer as an expert witness will be discussed in Chapter 6.0. The discussions include descriptions of various procedures for formal dispute resolution and the civil litigation process as well as descriptions of the forensic engineer's pre-trial responsibilities, trial preparation, testimony in court and post-trial consulting. The impact of being involved in a formal legal dispute, combined with the stress due to service as an expert witness, will be discussed in detail in Chapter 6. An engineer should carefully consider these factors before deciding to become an expert witness.

CHAPTER 2

THE INVESTIGATION

This Chapter describes the methodology for investigative procedures that may be used to identify the cause or causes of failure. The procedures and methodology outlined in this chapter are general in nature, and should be interfaced with specific investigative procedures as described in Chapter 4.0, "Geotechnical Failures" and Chapter 5.0, "Structural Failures". The forensic engineer may use these general procedures to investigate and report the cause of the failure and/or to evaluate the nature and extent of the failure. In addition, these concepts may be used to prepare studies for situations in other engineering disciplines that require techniques for investigation, analysis, development of conclusions, and preparation of comprehensive reports suitable as a legal instrument.

The methodology for a successful failure investigation includes a series of items, (i.e., procedures and operative tasks) that are detailed in this chapter. The discussion of each item includes its basic requirements, procedures for synthesis of the body of information developed during the investigation, as well as the development of hypotheses that will lead to an understanding of the probable causes of the failure.

2.1 PLANNING THE INVESTIGATION

The implementation of a successful investigation begins with the development of a logical investigative plan and establishment of the project goals. The forensic engineer will assume the role of team leader and should develop an investigation plan that will trigger a systematic process to acquire and analyze data, develop failure hypotheses and prepare the final report. The planning effort includes budget and scheduling considerations, selection of an interdisciplinary team that will best respond to the failure situation, site observations and testing requirements, document collection, analysis and synthesis of data, and development of failure hypotheses. Planning includes outlining the operative tasks and procedures that together comprise the methodology for failure investigation, as well as the budgetary and scheduling requirements for each phase. Following is an outline of the tasks which comprise the critical elements of the investigative plan. These tasks and procedures are discussed in the remainder of this chapter.

1. planning the investigation
2. client interface/project schedule/budget
3. identification of the investigative team
4. operations planning
5. site observations and analysis
6. document search
7. literature search
8. investigative synthesis
9. development of hypothesis

2.2   CLIENT INTERFACE/PROJECT SCHEDULE/BUDGET

When the investigative plan has been outlined, conferences should be held to apprise the client of the scope of work required for a proper investigation. As discussed in Chapter 1.0, this should be done in the form of a written proposal, along with an agreement for forensic engineering services. The investigation plan contained in the proposal should include an outline of the anticipated tasks and efforts required to conduct the investigation.

The budget included in the proposal should reflect the estimated cost of the operation, considering the anticipated schedule and the costs for the forensic engineer, staff, consultants, testing, travel, graphics, and printing.

The time schedule submitted in the proposal should extend from the initial conferences with the client through the completion of the report. It should define the time period required for the testing laboratories to complete their work and the time periods anticipated for document and literature search, the preparation of graphics and analysis, the synthesis of data, development of hypothesis and preparation of the report. The schedule and budget should be updated periodically as changes in the investigation plan or client requirements are altered. It should be made clear to the client that the scope of the investigation may change as new facts are uncovered.

The client's representative should be identified during this stage. It is important to establish a single individual who will represent the client in order to maintain a direct line of communication and to assure confidentiality.

2.3 IDENTIFICATION OF THE INVESTIGATIVE TEAM

When carrying out an investigative commission that involves a single discipline, the forensic engineer may require little or no assistance from other disciplines. However, when commissioned to carry out the investigation of large and/or complex failures, the forensic engineer will retain professionals from other disciplines while serving as principal investigator and team leader.

The individual forensic engineer rarely has sufficient expertise or in-house staff to deal with the interdisciplinary aspects of a complex investigation, and therefore must rely on consultants who are experts in other fields. The individual experts should be selected not only for their technical competence and ability to communicate, but also for their capabilities as expert witnesses since they may be called upon to testify.

The experienced forensic engineer should have a working relationship with a number of experts from other disciplines. Keeping an up-to-date file and occasional contact with these experts will facilitate their availability on short notice in the event their services are needed.

To conduct a comprehensive investigation, the forensic engineer as team leader may require experts from the following disciplines:

1. geotechnical engineering/testing
2. materials technology/testing
3. chemical technology/testing
4. metallurgy technology/testing
5. dynamics technology
6. acoustical/vibration engineering
7. construction management
8. construction scheduling
9. mechanical engineering
10. electrical engineering
11. structural engineering
12. civil engineering

13. surveying
14. photography/photogrametry
15. video camera operations
16. construction equipment operations
17. maintenance/operations experts
18. safety engineering

2.4 OPERATING PLANNING

In developing the operations plan required to
implement the investigation, the forensic engineer
in the role as principal investigator will be
involved with coordinating, analyzing and
integrating the work product of the team members.
An organizational plan should be developed to
enhance communication between the various team
members. Standard reporting measures should be
established.

Periodic team meetings should be held. At these
meetings, overall progress of the program and
intermediate results of the investigation are
reviewed, coordination of operation efforts
implemented, failure hypotheses discussed, and the
investigation plan refined or altered as
required.

The forensic engineer as principal investigator
should develop a project directory which will
involve the identities of experts on the team,
their telephone numbers, addresses and a
description of their respective roles in the
investigation.

2.5 SITE OBSERVATIONS/TESTING/ ANALYSIS

The forensic engineer should visit the site to
personally ascertain the initial conditions of the
failure or performance problems. The initial site
visit provides the engineer with personal
observations of the situation and the overview
necessary for an in-depth site analysis of the
incident. An expert with firsthand knowledge of
actual failure conditions has far more influence
than one whose information was garnered through
photographs or by other means. Several additional
site visits are usually required to carry out
specific investigations and testing programs.

The goals of site visits should be to conduct
overall visual examination, collect graphic and

narrative records, obtain eyewitness accounts, and perform testing programs. Chapters 4.0 and 5.0 will discuss specific tasks for observations and testing for geotechnical and structural failure investigations.

Following are methods and techniques of optimizing the data to be obtained from site visits.

2.5.1 INITIAL SITE VISIT

The purpose of the initial site is to evaluate the scope and nature of the incident of failure, and to aid in preparing the investigative plan, which includes the following:

1. an outline of the investigative plan
2. equipment required for site visits
3. field and laboratory testing procedures
4. development of the interdisciplinary team of experts required to properly execute the investigation

The initial site visit may be conducted by the principal investigator alone, or the principal investigator may be accompanied by an expert or experts who will be involved in the failure investigation when the nature of the failure is identified (e.g., a metallurgist if a steel structure failed), a photographer with the proper equipment for long-range or close-up photos of the site, and a technical record-keeper to maintain and collect initial information.

2.5.2 EQUIPMENT FOR SITE VISITS

The proper equipment for the initial site visit, as well as subsequent visits, will play an important role in the investigation team's ability to record the incident of failure.

Each team member must be responsible for maintaining that member's own equipment. The majority of the equipment can usually be stored in a shoulder bag between assignments. Examples of the possible content of this kit are listed below:

1. scale: 12" rule (.01" division), 6' tape, 100' tape
2. hand-held level
3. photographic equipment
4. video camera
5. communicator, such as a two-way radio
6. safety gear: hard hat, gloves, steel-toed boots, rainsuit, safety glasses
7. sample collection: tags, duct tape, plastic bags
8. lights: flashlight or penlight
9. recording materials: notebooks, graph paper, clipboard, pocket tape recorder
10. calculator
11. hand Tools: Mason's or geologist's hammer, pocket knife, pocket saw, ice pick
12. markers: waterproof felt tip, chalk, spray paint

In addition, if the failed facility is inaccessible (e.g., is elevated) the principal investigator should arrange for equipment that will enable access to the focus of the investigation. The equipment may be available from the contractor at the site, or equipment and operators may be available from local equipment rental companies. Equipment that can be used to provide safe access to the failure might include one of the following:

1. telescoping lift (snorkle lift)
2. cherry picker crane with personnel bucket
3. safety pipe scaffolding supported on ground or slab
4. hanging scaffold or stage
5. ladders

Each piece of equipment must be operated in a safe and secure fashion. Emotions run high at failure sites, and the principal investigator must assure that proper and safe procedures are used in operating the equipment. An experienced operator or erector must be involved and safety belts, lifelines, safety netting, and redundant methods of support should be used.

2.5.3 ARRIVAL AT THE SITE

The principal forensic investigator should arrive at the site, either before the team or along with the team. This investigator should present personal identification, identify the client (if required), and state the nature of the team's business at the site to the person or persons in charge. In the case of a catastrophe or a collapse, the organizations in charge may be the National Guard, local or state police, fire-fighters, or contractors. If identification checks or attendance lists are being implemented, then the principal investigator should assure that the investigator and the team are on the record as having been at the site.

2.5.4 SAFETY AND RESCUE OPERATIONS

Sometimes a failure causes death and/or injuries to persons who are trapped in the debris and rescue operations are threatened by further collapse of the damaged structure. In this event, it is likely that the principal investigator, as a professional engineer, may be the best qualified person at the site to evaluate the safety of a partially standing structure.

The qualified professional who renders assistance in safety/rescue operations must carefully consider the life safety and liability consequences of his or her actions. The engineer who engages in failure investigations should be aware of the possibility of being in a position to offer experience based on engineering judgment and advice in adverse conditions (e.g., a partially collapsed structure) without benefit of calculations. Each professional should establish their own policy for dealing with such circumstances.

2.5.5 REMOVAL OF DEBRIS

In the event that rescue operations are required to remove debris from the original in-place failure position, or, in the event that clean-up operations have begun prior to the arrival of the investigator at the site, the investigator should obtain graphic

images of the initial failure scene,
including overall photographs, sketches, and
on-the-spot interviews with the clean-up
crew. Interviews with rescue crews also may
provide information otherwise lost or
forgotten.

It would be desirable that debris not be
removed until the principal investigator has
had an opportunity to photograph and study
the debris in place, and collect the
necessary data and specimens. This, of
course, is not possible when human life is
threatened.

The investigation team should record
significant data as the debris is removed.
Each piece should be categorized, photo-
graphed in place and sketched, along with
its proximity to other pieces, and stored in
a systematic fashion. This can be
accomplished in a number of ways, but the
objectives for a recording and storage
program of failure debris are as follows:

2.5.5.1 Orientation of Debris

It is desirable, through the use of
sketches and photographs, to create
a record of the position and
orientation of each section of the
facility after the collapse.
Three-dimensional sketches generated
by hand at the site are important.
These sketches and the accompanying
notations provide a perspective of
the investigation that cannot be
determined from photographs. In
addition, they are invaluable in the
event the photos are lost or are of
low quality.

2.5.5.2 Storage of Debris

The principal investigator should
attempt to find a convenient and
secure location where the failed
elements of the facility may be
stored after removal from the
failure site. The storage area
should permit examination and study

of the debris in a protected environment.

2.5.5.3 Reconstruction of Elements

The retrieved elements may be laid out in an orderly fashion in the "before collapse" geometry. This technique will aid in recreating the movement of the failure and the sequence of events that transpired prior to and during the collapse.

2.5.5.4 Source for Testing of Materials

The removal and storage of debris to a secure site will provide a protected source for sampling of the materials of construction as well as for field and laboratory testing of the engineered systems.

2.5.6 OVERALL VISUAL OBSERVATIONS

The initial examination of a failed facility should take place upon arrival at the site. This important task should be performed, no matter how briefly, by the principal investigator. Refer to Chapter 4.0, "Geotechnical Failures" and Chapter 5.0, "Structural Failures" for salient descriptions of specific items to be recorded during the initial visual examination.

2.5.7 DOCUMENTATION OF VISUAL EXAMINATION

The compilation of graphic and narrative records of the investigation should be implemented in order to provide an in-depth history of the failure scenario. Field data can be collected in the form of sketches, photographs, video recordings, verbal descriptions and eye-witness reports. Following are descriptions of some techniques for data collection that can be accomplished through the use of the above mentioned media.

2.5.7.1 Sketches

The use of well-conceived and competent sketches are vital to the investigation. Sketches of the overall configuration and detailed sketches of specific areas will augment other types of data collection at the site. The very act of preparing a sketch may jog the investigator's memory, thereby helping him or her identify the failure because of an association or comparison with similar past failures. Furthermore, transcribing sketches will reinforce the forensic engineer's memory of the incident and its minute details.

2.5.7.1.1 Reference System

As part of the investigation plan, a reference system of coordinates should be established and used for documentation of the location of the total facility and the spatial relationship of the distressed parts of the facility. Sketches should always include the location of the subject area using the reference system. If the construction plans are available, the building coordinate system can be used. The reference system should also be used when identifying the contents of photographs as well as in verbal and written reports and eyewitness accounts.

2.5.7.2 Scale and Orientation

To aid in reading the sketches at a later date in the investigation, it is recommended that a pad of sketch paper with a grid be utilized in order to facilitate orientation and assist in maintaining the scale of the sketch. It is important that all sketches be drawn as near to scale as possible.

2.5.7.3 Types of Sketches

Sketches should be made of large areas as well as detailed views of the failure. The investigation team should prepare overview sketches which graphically indicate the configuration of the failure in relation to the original configuration as well as to the remaining portions of the facility. This type of sketch should be rendered in three-dimensions and should indicate the investigator's assumptions and, if possible, thoughts pertaining to the mode and direction of failure. The sketch should be annotated with comments and vectors indicating failure motion.

2.5.7.4 Photographs

Photography as a tool for collecting data is valuable during the initial site visit, and afterward for use as a visual recording instrument in the on-going effort of data collection.

During the initial visit, numerous photographs should be taken from various angles. The photograph should be all-encompassing, concentrating not only on the area of failure but the entire facility. The use of multiple wide-view and detailed photos will record data and situations that may be removed or disturbed at a later date if the area is found to be germane to the investigation.

It is not usually necessary to utilize complex photographic equipment for the all-encompassing photos; a good fixed lens 35 mm camera with a built-in flash will facilitate rapid photo taking.

Professional photographers can provide photographic expertise and

permit the investigation team freedom
to concentrate on orientation as well
as the framing and logging of photos.

Cross-referencing sketches prepared
at the site with camera orientation
can be of great assistance in the
analysis of the failure.

Aerial photographs of large scale
failures can provide good perspective
for overall movements and effects,
and can be used to orient portions of
the failed facility to the original
location.

Photography can sometimes be used to
identify overall distortions and deflections not detectable by other
means.

2.5.7.5 Videotaping

Videotaping with audio backgrounds
will provide a visual and narrative
overview of the failure site. The
videotape may be used during analysis
of the failure and later during the
legal process in court proceedings.
The use of videotaping should follow
the procedures described for
photographs.

2.5.7.6 Verbal Descriptions

One of the most powerful techniques
for acquiring data is recording
impressions while examining the
failure. An effective way to do this
is with a pocket recorder.
Succinctly record thoughts or
impressions as they occur. Later,
transcribe , edit and review the
thoughts in a controlled environment.
This provides an opportunity for
examination and consideration.

2.5.7.7 Eyewitness Interviews

The use of eyewitness accounts will
provide background data and insight
into the cause, mode, and sequence of

-34-

the failure. Conversations with persons who were present before and/or during a failure can be interfaced with the other data and documentation generated during the investigation to develop an overview of the failure scenario.

Interviews should be conducted with the operators of a facility, construction workers, users, tenants, passersby, public safety officials, etc.

A thorough job of interviewing eyewitnesses should present a verbal description of the condition of the collapsed structure before, during, and after the failure. It is anticipated that the usual document retrieval effort will produce printed documentation that records the structure's history. However, interviewing people about the structure's past behavior is necessary if the investigation team cannot find written documentation.

Interviews should contain eyewitness accounts of the facility and the failure through questions relating to prior events, sensory responses (e.g., impressions of odors, vibrations, visual movements, failure of non-structural components, or sounds associated with the failure). Since this part of data collection is likely to reveal information of a non-technical nature, it may not be necessary for the persons conducting the interviews to have technical backgrounds. It is advantageous to select interviewers who possess the following attributes:

1. Gregariousness and the ability to communicate easily with persons of widely varying backgrounds.

2. A native of the area near the collapse.

3. The ability to conduct a structured interview.

4. The ability to follow lines of thought that generate new and meaningful data.

5. The ability to ask incisive questions that can separate fact from fantasy.

The format of the interview should be developed so that the interviews are consistent, reliable, and thorough. The interview should be conducted so that each person interviewed is allowed to contribute all the pertinent information he or she possesses. The interview can vary to some extent, depending on the scope of the project, its use, age, and performance; but once an interview structure is established by the lead investigator, it should be maintained throughout the investigation unless altered by the lead investigator.

The interviewer should be sensitive to the feelings of the person being interviewed. This is especially true if the person being interviewed, or his or her relatives or friends, were directly involved or affected by the failure. An on-site interview, although preferrable, may not be as productive as a later interview, especially if adverse conditions of any sort exist.

To initiate the interview, the interviewer should first introduce himself or herself, show some identification, and then ask the person to make an appointment if he or she is not immediately available. Once the interview begins, it should include inquiries regarding all the subjects listed on an Eyewitness Interview Form. After completing the interview, each interviewer should record his or her impressions of the eyewitness's acuity, eye for detail, and general reliability. The answer to Question No. 16 in the following format may provide information as to whether the eyewitness is prejudiced.

Eyewitness accounts should be evaluated immediately by the principal investigator and the investigator's comments should be recorded directly on the interview form. It should be anticipated that conflicts and inconsistencies will be detected among the various eyewitness accounts. The chief investigator working with the team must sort these out, and a second round of interviews may be necessary to resolve conflicting statements.

Eyewitness Interview Format

1. date of interview

2. person's name, address, age, occupation, affiliation with the project

3. knowledge of site prior to construction

4. knowledge of site during construction: dates of construction, names, addresses, and phone numbers of persons who constructed the project

5. direct knowledge of the use of the facility

6. description of eyewitness's opinion (prior to failure) about use of the structure

7. description of occurrences the day of the failure

8. description of occurrences the day or days prior to the failure

9. where eyewitness was at time of failure

10. description of failure and its sequence

11. description of eyewitness's vehicle

12. description of occurrences immediately after failure

13. description of facility after failure

14. opinion of clean-up operations

15. how the failure affects witnesses, their families, and friends

16. opinion of the cause of failure

2.5.8 TESTING PROGRAMS

The forensic engineer and the investigation team will be responsible for the development and implementation of testing programs for the distressed facility, including acquisition of test specimens that will be used to develop background characteristics of the materials of construction and/or assemblies. The testing programs may include physical and chemical tests on materials of construction, testing of the various elements of equipment, and load tests. Some tests may be conducted in the field. However, the majority of tests will be carried out in the testing laboratory using samples taken in the field. In addition, the engineer is responsible for the continuous care and custody of test specimens.

The methodology for preparing field and laboratory testing programs and discussions of specific applications are contained in Chapter 4.0, "Investigations of Geotechnical Failures" and Chapter 5.0, "Investigations of Structural Failures".

2.5.9 ORGANIZATION OF FIELD NOTES AND DATA

The neatness, organization, and completeness of notes, sketches, and logs for the investigation effort are important, since the forensic engineer may need to refer to the notes years after the failure if the incident results in litigation. In addition, the forensic engineer, when serving as expert witness, will need to refer to the notes in court or deposition, in which case the notes will become exhibits subject to close scrutiny by hostile parties.

Clarity will assist the expert in recalling the incidents recorded in the notes. Thus the data generated during the investigation should be organized in notebooks and/or files with a clearly defined system of nomenclature.

2.6 DOCUMENT SEARCH

The forensic engineering team will be responsible for the acquisition and review of all available documents relating to the design and construction of the facility.

The body of documents that are generated during the development of a project should be reviewed to provide an overview of the history of the project. It should also be analyzed to assess and identify potential technical or procedural errors. The project documents provide specific information about:

1. The history of the development, design, and construction of the facility.
2. The condition of the facility at the time of failure relative to maintenance and/or alterations.
3. The use of the facility at the time of failure relative to its actual loading compared with the original service loads employed for design.

The foregoing can be interfaced with additional analyses in order to identify and allocate responsibility for the causes of the failure.

In certain situations all parties in a legal dispute will agree to make all documents in their files available for use by all other parties. In some situations, such as large law suits, the documents may be turned over to a central depository for use by various parties.

However, the client's attorney will often request specific documents from the various parties through the legal process known as "discovery." In order to assure the systematic collection of all documents germane to the investigation, the forensic engineer should prepare a list to be used by the client's attorney when requesting the documents. In order to assist the forensic investigator in developing the listing of desired documents, this section provides an overview of typical documents that are generated during the development of a project and used in the investigation process.

2.6.1 CONTRACT DOCUMENTS

The contract documents include the contract design drawings, contract specifications, construction contracts, general conditions to the contract, special and supplementary conditions, field change records and change orders, information bulletins, and shop drawings. Following is a checklist of the contract documents:

2.6.1.1 Contract Design Drawings

1. Architectural
2. Structural Engineering
3. Civil Engineering
4. Mechanical Engineering
5. Electrical Engineering

2.6.1.2 Contract Specifications

1. General Conditions
2. Special Conditions
3. Supplementary Conditions
4. Technical Sections of the Specifications

2.6.1.3 Contracts (Contractual Agreements)

1. Owner/Architect Contract
2. Engineer/Architect Contract
3. Owner/General Contractor Contract

2.6.1.4 Contract Provisions

1. Addenda to the Construction Documents
2. Information Bulletins

3. Field Directives
4. Change Orders to the Contract
5. Correspondence authorizing changes to the Contract Documents

2.6.1.5 Shop Drawings

Shop drawings are drawings that are
prepared by the contractor to indicate in exact detail the fabrication
and erection of the components of the
project:

1. Architectural shop drawings
2. structural engineering shop
   drawings
3. mechanical engineering shop
   drawings
4. electrical engineering shop
   drawings
5. civil   engineering   shop
   drawings

2.6.1.6 Project Payment Certificates

The project payment certificates are
generated by the general contractor
each month and indicate the progress
of work and the dollar amount requested and paid. These also indicate the amounts paid to each of the
major subcontractors on the project.

2.6.1.7 As-Built Drawings

The preparation of as-built drawings
are not the rule on the majority of
projects. As-built drawings may be
prepared by the design/construction
team using several alternate forms,
which include:

1. Altering the original repro-
   ducible documents to indicate
   changes in the original design
   during the construction of the
   project.

2. A record set of prints of the
   original construction documents
   with attached copies of sketches

that indicate changes to the original design, or with changes noted directly on the prints.

2.6.2 TEST REPORTS/MILL CERTIFICATES

The test reports generated during the design and construction of a project are the documents that describe the quality of the materials of construction, building systems, equipment, installation procedures, or site/utility conditions. These reports generally include the following:

2.6.2.1 Testing Laboratory Reports

1. Foundation Report: Subsurface analysis
2. Wind Tunnel Tests: Reports relative to wind tunnel tests on a scale model of project
3. Soil Reports: Compaction testing, soil bearing or rock bearing values taken during construction
4. Structural Steel Testing: Tests on welded or bolted connections and on the material properties
5. Concrete Material Testing: Tests for compression strength, grout cube testing, tensile splitting strength, and tests for the modulus of elasticity
6. Concrete Test Mixes: Reports indicating the quantity and quality of materials to be used in the concrete and tests on samples of this concrete prior to production
7. Certification of Welders: Verification of capabilities of welder by acceptable standards
8. Structural Load Tests: Reports on testing of structural components
9. Post-Tensioning Reports: Elongation of tendons, etc.
10. Pressure Tests: Testing of pressure of water lines
11. Flow Tests: Tests for flow in sewer lines

2.6.2.2 Mill or Manufacturing Certificates

1. Structural Steel Mill Certificates
2. Reinforcing Steel Mill Certificates
3. Post-Tensioning Tendons Mill Certificates
4. Portland Cement Mill Certificates
5. Concrete Masonry Unit Strength Certificates
6. Welding Electrodes Certification

2.6.3 FIELD REPORTS

Field reports are generated by various parties responsible for the design and construction of the project. Reports may be prepared daily, as in the case of the contractor's superintendent, or a full time clerk of the works, representing the owner; or they may be prepared periodically, as in the case of field reports from the architect or engineer.

1. Clerk of the Works: Daily reports from the architect's or owner's full time representative
2. Resident Engineers: Daily reports from the full time representative of the engineer/design firm
3. Construction Manager: Daily logs and reports
4. Construction Superintendent: Daily logs
5. Engineering Field Reports: Periodic reports prepared by engineers of record, including civil, structural, mechanical, and electrical engineering
6. Architect's Field Reports: Periodic reports prepared by the architect's representative

2.6.4 INSPECTION REPORTS

Inspection reports are generated by professionals who are not members of the design and construction team. The reports are generally prepared by governmental agencies or lending institutions that periodically report on the quality and progress of the work.

1. Building Inspection Reports: Reports generated by local building authorities
2. Lending Institution Reports: Reports prepared by independent engineering firms retained by construction and permanent lenders
3. Owner's Inspection Reports: Reports generated by independent engineers retained by the Owner.

2.6.5 PROJECT SCHEDULES

The project schedules are developed over the life of the project. They are usually initiated during the feasibility phase of the project and extend through the completion of construction. Schedules may be developed in various formats, including narrative schedules, bar charts, and CPM or PERT schedules.

1. Design Schedules: Schedules generated by the Architect, Engineer or Construction manager indicating start and completion of various phases of design and the interface of design activities with construction.

2. Construction Schedules: Schedules and periodic updates generated by the construction managers or general contractors that indicate construction progress and delays from inception to completion.

3. Design/Construction Schedules: Schedules generated by the owner, construction manager or general contractor that indicate the relationship between design activities and construction activities, generally found in a fast track or design/build project.

2.6.6 PROJECT CORRESPONDENCE

Project correspondence between the various parties should be collected and should include items generated from the inception of the project through the post-construction period:

-44-

1. Owner/Developer Correspondence
2. Owner/Consultant Correspondence
3. Owner/Contractor Correspondence
4. Consultant/Contractor Correspondence: Architect, Engineers and Construction Managers
5. Intraconsultant Correspondence: Architect/Engineer
6. Transmittal Records: All parties
7. Memoranda: All parties
8. Minutes of Meetings for Pre-Design and Design Period
9. Telephone Records: All parties
10. Minutes of Pre-Bid, Pre-construction and Construction Progress Meetings

2.6.7 CONSULTANT REPORTS

Consultant reports are prepared by professionals with special expertise who supplement the input of the architect and engineer of record. These reports are usually prepared prior to the initiation of the design phase and reflect on the viability of the project.

1. Traffic Studies: Reports relating to pre-existing and final configuration of traffic at project site
2. Planning Reports: Report indicating demographics and the Master Plan for project
3. Feasibility Studies: Marketing and economic studies, usually prepared by a firm of certified public accountants
4. Design Studies: Design alternatives prepared by the architect to determine the optimum scope of project, may also consider alternate site locations
5. Scheduling/Progress Reports: Reports generated by the consultants including initial schedules and periodic updating of schedules. These reports also indicate construction forecasting and progress.
6. Geotechnical Studies: Studies indicating the subsurface exploration, program analysis and recommendations for foundation design
7. Utility Studies: Reports indicating location and capacities of utilities, flow studies on sewage system, pressure

tests on water system, and location of gas and electric service.

2.6.8 WEATHER RECORDS

Weather records may be obtained from the closest United States government weather reporting station. Records may be requested for the time period extending from the date of construction to the date of the failure incident. Weather data available includes temperature, humidity, precipitation quantity and type, velocity and direction of wind, and atmospheric turbidity. This data may be obtained from the National Oceanic and Atmospheric Administration (NOAA), National Climatic Center, Asheville, North Carolina.

2.6.9 DESIGN ANALYSIS

Collection and documentation of the studies and design calculations carried out by the architect, engineers, and special consultants will provide an overview of the evolution of the design process. The body of narrative and mathematical calculations and graphics prepared during the progress of the design of the facility should be collected.

1. Building Code: Acquire a copy of the Building Code in force during the design of the project.
2. Engineering Analysis: Calculations, graphics, and studies prepared by the civil, structural, mechanical, and electrical engineering disciplines.
3. Architectural Analysis: Graphics, calculations, design alternatives, studies and notes prepared by the Architect of Record in developing the design for the project.
4. Design-Build Analysis: Graphics, calculations, notes and studies prepared by subcontractors on a design-build basis, including post-tensioning design, precast concrete documents, earthwork and shoring, mechanical systems, and electrical systems.
5. Special Engineering Studies: Calculations and studies prepared by special consultants for vertical transportation, acoutics, off-site hydrology, etc.

2.6.10 MAINTENANCE AND MODIFICATION RECORDS

The document search should include the collection of maintenance and post-construction modification records of the building and equipment. These documents could provide information with regard to components of the projects that have experienced a failure during the service life, as well as components or details that have been modified or are being used for applications other than that intended by the original design and construction.

2.6.11 BUILDING DEPARTMENT PERMITS

The building permit documents that are required by the local government agencies will provide information pertaining to the design, construction, and development of the project. The document search should include the records filed with the Building Department when the permits were granted. These documents normally include statements of ownership, the design team, contractors, cost data, and the set of contract documents approved for construction.

The following permits should be researched:

1. building permits
2. foundations permits
3. site development permits
4. water/sewer tap permits
5. certificate of occupancy

2.7 HISTORICAL INFORMATION/VISUAL DOCUMENTATION

In certain failure situations, the investigative team may not gain access to the site for some time after the failure. Depending on the situation, evidence may have been removed and/or repair procedures may have been implemented. It is possible to collect historical data from various sources in order to develop a body of information that recreates the condition prevailing at the time of the failure, as well as the configuration and condition of the facility prior to the failure.

The best sources of visual information, including photographs and video recordings, are often the media, local governments, and insurance companies (among others). Following is a check list of potential sources of graphic evidence that may be used by the forensic team when investigating a failure:

1. television: Networks and local stations
2. newspaper files
3. insurance adjusters
4. building owners or operators
5. police departments
6. fire departments
7. the Civil Defense rescue team
8. local building inspectors
9. neighbors

2.8   LITERATURE SEARCH

An extensive body of published works exists dealing with historical accounts of the failures of engineered facilities, techniques for investigating failures, and methodologies for the prevention of failures. The collection and review of published works relating to the subject failure will provide background data on the failure; assist in developing a failure profile based on similar failures, and assist in developing failure hypotheses.

This guide includes a bibliography that represents an overview of literature relating to civil engineering facilities. The bibliography is intended to assist in the initiation of a comprehensive literature search.

In addition to this bibliography, other possible sources for literature include:

1. Professional Societies: The publications of professional engineering societies may be contacted for assistance in providing bibliographies relating to engineering failures.

2. Trade Associations: Trade associations publish magazines, journals and newsletters that contain articles on engineering failures. These organizations publish indexes to their own publications, which are useful in a literature search.

3. Proceedings of Conferences and Symposiums: Publications of the proceedings relating to failure investigations can be researched.

4. Engineering Libraries: Libraries in most universities contain bibliographies and literature relating to failures.

5. Computer Data Bases: Many universities and certain commercial sources have on-line services that provide literature searches for publications relating to failures.

The bibliography at the end of this guide lists a great deal of reference material that range from textbooks to articles in trade publications. In order to facilitate the literature search, the bibliography has been sub-divided into several categories, including:

1. General: Titles relating to aspects of general investigations.

2. Legal: Literature relating to legal aspects encountered by the expert witness.

3. Structural: Literature relating to structural failure investigation.

4. Geotechnical: Literature relating to geotechnical failure investigation.

2.9 CLASSIFICATION OF FAILURES

Webster's Unabridged Dictionary describes a failure as follows: **a falling short, deficiency or total defect.** In reality a failure does not necessarily mean total collapse, but can equally apply to a shortfall, such as when a constructed facility does not perform as it was originally intended. The identification of a failure and its classification according to type, cause and time phase could be the first step in the identification of the cause of the failure.

The failure can be classified in terms of the following criteria:

1. time: age of the facility with respect to its service life
2. type: the extent of the failure

3. cause: incidents causing the failure

Classification not only assists in establishing the type of failure, but helps determine the timing of events for use in the investigation, analyses, and preparation of the report.

The investigation of a failure should be based on a systematic analysis that includes the identification of the physical as well as the technical and procedural causes of the incident. The following definitions apply to the three specific categories that can be used to identify and classify a specific failure: the time phase, the type of failure, and the cause of failure.

2.9.1 TIME PHASE

The classification of a failure with respect to time is based on the intended service life of the original construction.

The service life of a facility is the time period during which the constructed facility is in use and functions for the purposes intended by the original design criteria. The service life of a facility starts with the completion of initial construction and extends until its demolition.

Following are definitions of the three basic time phases during which the incident of failure could occur:

2.9.1.1 Pre-service Period

Failures occurring during the construction period or any time prior to initiation of the intended service usage.

2.9.1.2 Service Period

Failure occurring during the service life for which the construction was designed and built.

2.9.1.3 Post-service Period

The post-service period is initiated when the project is used for a purpose other than the intent dictated by the original design criteria. Failures during the postservice period are generally due to overloading or degradation. Failures during demolition are included in this classification.

2.9.2 TYPES OF FAILURES

The identification of the type of failure relates to the extent and nature of losses (e.g., human life or economic) due to the failure. The classification of a failure with respect to the extent and type of damage can be identified as follows:

2.9.2.1 Safety Failures

This type of failure involves safety and is the result of either the total collapse or the partial failure of an engineered facility, and results in death or injury, or failures that place human life in jeopardy. This type of failure gains public attention because of loss of life, and/or injuries, and often results in great economic cost. Included in this category are building collapses, dam failures, or fire safety failures.

2.9.2.2 Functional Failures

Functional failures are the most common type of failures. They impair the normal use of an engineered facility, have a negative impact on the serviceability of a facility, and compromise the expected usage of a facility.

Some examples of functional failures are:

1. water penetration failures
2. joint movement failures
3. mechanical system failures
4. foundation failures (differential)

5.   excessive building sway
     (horizontal deflection)
6.   aesthetics
7.   high BOD or effluents
8.   highway potholes
9.   stream bank erosion
10. excessive deflections
11. unacceptable vibrations
12. acoustical problems
13. premature material deteri-
     oration

2.9.2.3 Latent Failures

This type of failure refers to
situations where collapse has not
occurred, but collapse is imminent
due to an undetected weakness in the
engineered system. This type of
failure may be detected by an
engineering review, testing of
substandard construction material,
load testing, or inspection of a
substandard or weakened construction detail. A latent failure is a
problem waiting to happen.

2.9.2.4 Ancillary Failures

Ancillary failures do not impinge on
the safety or the function of an
engineered system, but deal with the
alteration or extension of the time
of construction or the escalation of
construction costs for an engineered
construction. This type of failure
includes overruns in construction
costs or schedules.

2.9.3 CAUSES OF FAILURES

The cause of a failure is the incident or
series of incidents that directly caused
the failure. Causes of failures can be
categorized into two classifications:
technical and procedural.

2.9.3.1 Failures Due to Technical Errors

Failures caused by errors in the
execution of the engineering design
or construction of an engineered

construction can be identified as
technical errors. Following are examples of technical errors:

2.9.3.1.1 Failures Due to Design Errors (Omissions)

This type of failure stems
from an error in the en-
gineering design of a
facility. The mechanism
causing the failure could
be the result of a design
error wherein the building
component or mechanism is
improperly sized or developed, or the failure could
be caused by the omission
of a critical part of the
mechanism or system.

The inclusion of an er-
roneous design, miscal-
culation, or omission in
the contract documents
could cause a variety of
failures ranging from a
nuisance to a total
collapse. This type of
cause could be the direct
result of lack of
experience, negligence,
lack of education, incom-
petence, or the inability
to communicate.

2.9.3.1.2 Failures Due to Detailing or Specification Errors

Failures resulting from
detailing errors on the
contract drawings or errors
in the preparation of
project specifications can
contribute to a failure
situation.

The inclusion of misin-
formation, incorrect data,
improper or erroneous
materials, equipment, or

techniques can lead to a variety of failures.

2.9.3.1.3 Failures Due to Construction Errors

Failures caused directly by omission of critical details or the comission of an error during the construction process can lead to a variety of failures ranging from a nuisance failure to a complete collapse.

Failures involving construction errors can be due to the omission of critical system components: utilization of substandard materials or equipment, improper construction processes, out-of-sequence installation of details and utilization of out-of-alignment construction, or lack of quality control.

2.9.3.1.4 Failures Due to Deficiencies in Materials of Construction

This type of failure is due to errors in the application of construction materials during the construction process. Failures from deficiencies of materials can be caused by installation during adverse weather conditions and by construction practices that reduce the strength or durability of materials due to field tampering.

2.9.3.2 Procedural Errors

Failures caused by procedural errors involve incidents which are the result of miscommunication, out-of-sequence operations, mixed sequences, lack of coordination, excessive speed or emphasis on economy, regulatory influence, poor response time, and improper attitude.

2.9.3.2.1 <u>Failures Due to Poor Coordination of Contract Documents</u>

These failures are caused by errors due to the lack of coordination between the various disciplines involved in the execution of contract documents, or by conflicts in information between the contract drawings and the specifications.

2.9.3.2.2 <u>Failures Due to Errors in Shop Drawing Review</u>

Failures may be caused by errors involving components, materials, or details that are specified correctly on the contract documents but erroneously detailed on the shop drawings. The error may not be detected by the reviewer during the shop drawing review process, and may not result in failure.

2.9.3.2.3 <u>Errors Due to Ineffective Coordination of Construction</u>

The lack of coordination between trades during the construction process, or lack of coordination between shop drawings and the construction process often leads to a failure.

2.9.3.2.4 Errors Due to Poor Communication

Failures often result from poor communication between various design disciplines, construction trades, or between the contractor and design disciplines. These "sins of omission" can lead to a falling short, a weakening or a total defect, or a cessation of proper functioning or performance.

2.10 SYNTHESIS OF THE INVESTIGATION

As data, documents and results are collected from the document search, literature, search, site observations, testing programs and design analysis, it is necessary to synthesize the findings of the investigation.

The principal investigator should work closely with the team to synthesize the efforts of the investigation and promote the intellectual process that will establish the most likely causes of the failure. The synthesis of the findings of the investigation will culminate with the development of failure hypotheses. Following is a description of the general methodology for the systematic appraisal of documents and data that will lead to the efficient development of failure hypotheses.

2.10.1 DEVELOPMENT OF THE HISTORY OF DESIGN AND CONSTRUCTION OF THE FACILITY

To provide an overview of the chronological milestones of the project, up to and including the incident of failure, a historical description should be developed using documents collected according to Section 2.6, including the following information:

2.10.1.1 History of the Failure Incident

Provides an overall description of the failure incident, including salient data and known information about the events before and during the collapse, and damage or disruption of services.

2.10.1.2 Construction History

Provides information describing the initiation and completion of the construction, plus information with respect to aberrant incidents during construction.

2.10.1.3 Project Directory

Provides a listing of firms involved in the development, design and construction of the project, including key staff members, their roles, addresses and telephone numbers.

Owner: Name and address of original owner and subsequent owners

Developer: Name and address of developer and listing of original staff

Design team: Name and address of architectural and engineering firms and key staff members

General
Contractor: Name and address of firm and their key personnel

Subcon-
tractors: Name and address of principal subcontractors and key personnel

Lenders: Name and address of construction and permanent lender and staff members

2.10.2   **REVIEW OF SITE AND SERVICE CONDITIONS**

Develop accounts of the site and service conditions prevailing at the time of the failure, including climatic conditions, service usage and traffic activities, and other pertinent information:

2.10.2.1   **Climatic Records**

Records of rain, snow, ice, wind velocity, and temperature.

2.10.2.2   **Service Loading**

Usage of the facility or equipment at the time of failure and comparisons with design intent or original design criteria.

2.10.2.3   **Traffic Volume**

Records of traffic flow and size of vehicles to relate to load cycles and vibration.

2.10.2.4   **Eyewitness Accounts**

Accounts of eyewitnesses observing unusual activities at the site or adjacent properties, such as construction blasting, low flying aircraft, etc.

2.10.2.5   **Material Hazards**

A history of past events such as floods, fires, landslides, overloading, temperature extremes, etc., that may have affected the facility or equipment.

2.10.2.6   **Site Conditions**

Assessment of site conditions with respect to previous construction, topography, subsurface geologic conditions, seismic status, utilities and other pre-existing site characteristics. These elements should be interfaced with site design criteria to ascertain validity of criteria used in the original design assumptions.

2.10.3 DESCRIPTION OF THE FAILURE

To provide an in-depth description of the failure, it will be necessary to utilize the field observation data, eyewitness accounts, and test results. Following is a methodology for synthesis of the failure description:

2.10.3.1 Graphic Analysis

Review and analyze sketches, photos, and videotaping to develop the disposition of the failed facilities or equipment.

2.10.3.2 Description of Components

Evaluate information related to the condition of components of the failed facility or equipment.

2.10.3.3 Materials of Construction

Evaluate the materials of construction of the facility or equipment to reflect their condition versus their design requirements.

2.10.3.4 Evidence of Degradation

Evaluate the components of the facility or equipment to assess degradation due to effects of the environment or corrosive materials.

2.10.4 REVIEW OF THE ORIGINAL DESIGN

Independent design analysis of the original facility should be conducted to ascertain the ability of the failed facility to satisfy building codes and service requirements. The review will identify incidents of errors or omissions in the design documents.

2.10.5   REVIEW OF THE AS-BUILT DRAWINGS

An independent analysis of the design as altered during the construction period should be undertaken to ascertain the ability of the failed construction to satisfy building codes and service requirements; this data can also be used in comparison with the original design intentions.

2.10.6   REVIEW OF THE MODIFIED DESIGN/SERVICE LOADING

An independent design analysis of modifications to the facility after completion of construction and/or changes in the service usage should be carried out to ascertain the ability of the failed construction to satisfy building code requirements.

2.10.7   ANALYSIS OF DOCUMENTS RELATIVE TO POTENTIAL PROCEDURAL ERRORS

Analysis of relevant documents, including construction contract documents, shop drawings, and field bulletins should be conducted to ascertain if procedural errors occurred. Procedural errors are described in Section 2.9.

2.10.8   CLASSIFICATION OF THE FAILURE

Utilizing the methodology described in Section 2.9, the failure can be classified with respect to type, cause, and time.

2.11 DEVELOPMENT OF A FAILURE PROFILE

A failure profile can be developed using the methodology for synthesis of data, described in Section 2.10, combined with historical information researched in applicable literature, and the experience of the investigative team. This failure profile will describe the type and nature of the failure, and will provide salient information that should lead to the development of failure hypotheses. Knowledge of past failure causes would establish a minimum listing of failure hypotheses for consideration that should be supplemented as the investigation progresses.

-60-

The failure profile will include the classification of the failure, the project type, and
the identification of errors including those due
to design errors, construction errors, materials
of construction, procedural errors, environmental
or service loading factors, or other reasons.

Specific failure profiles for structural and
geotechnical failures are described in Section
4.5, "Profiles of Geotechnical Failures", and in
Section 5.8, "Structural Failures."

2.12 DEVELOPMENT OF FAILURE HYPOTHESIS

The process of developing failure hypotheses varies with the complexity of the failure situation. During an investigation of a minor nature the forensic engineer might develop failure hypotheses working alone. However, during the investigation of a large-scale failure requiring a team of investigators, the development of failure hypotheses should involve discussions and conferences during several phases of the investigation plan. Thus, the development of a final failure hypothesis should be an evolutionary process with input from the entire investigation team.

Often after the initial site visit, a failure hypothesis may be suggested; then as data is collected and analyzed, the initial hypothesis may be proven, abandoned, or revised, based on the synthesis of investigative data. If the initial hypothesis is abandoned and a new hypothesis is generated, then additional site investigation, document review, and synthesis of the analysis process may be required until a final hypothesis is established.

2.12.1 METHODOLOGY FOR DEVELOPING FAILURE HYPOTHESIS

The forensic engineer is responsible for the synthesis of data and the development of failure hypotheses in order to determine the probable cause of the failure. To accomplish this goal, the forensic engineer should be responsible for assessment and analysis of field observations, testing data and documents, failure profiles, and design analysis. Chapter 2.0, "The Investigation" describes methodologies for analysis and development of failure hypothesis, and Chapters 4.0 and 5.0 describe specific failure investigations.

In developing and refining hypotheses, the focus of the investigation should not be narrowed too early in the effort, and in particular, the investigation should not be limited to apparently obvious causes of failure. A broad listing of potential failure causes should be established to keep all avenues open until each of the "pretenders" can be eliminated by positive proof.

The principal investigator must keep an open mind to new and plausible hypotheses at any stage of the investigation, even though it may require new avenues of investigation.

When brain-storming to develop hypotheses for a failure it is advisable to use creativity. Creativity does not mean "reinventing the wheel," but simply developing several alternative hypotheses.

It is possible that after synthesizing the data of the investigative effort, differences of opinion will exist among the team members as to the primary cause of the failure. Complex failures are often determined by several factors acting together cumulatively. While the primary task of an investigation should be to determine the probable cause of the failure, it is possible to present findings which list several possible causes. In that event, an attempt to determine the primary and secondary causes of the failure should be made. Generally, the lead investigator is asked to determine the "most probable" cause.

CHAPTER 3

THE REPORT

The reports generated during a failure investigation should exemplify professionalism and the
high technical standards that are the hallmarks of
a properly executed investigation.

The various reports that chronicle a forensic
investigation should encompass the complete
history of that fact-finding mission. They should
begin with the failure incident, traverse the
complete investigation, and provide accounts of
critical events. The reports will be key
instruments in the legal action that usually
follows the investigation.

This section will discuss the types of reports
that might be generated during the investigative
process as well as provide a rationale and
guidelines for the production of a clear and
concise document.

3.1 TYPES OF REPORTS

Depending on the character of an investigation,
several different types of reports may be
requested and produced during the course of the
effort. Following are the types of reports that
could be encountered during an investigation.

3.1.1 THE ORAL REPORT

Oral reporting of a failure investigation
will usually precede any written report, or
may be requested in lieu of a written
report. The oral report will be given
directly to the client or the client's
attorney in order to inform them of the
progress of the investigation. The oral
report should be delivered after the
initial review of the failure. The
contents of the report should clarify the
investigator's understanding of the
assignment's scope and should include the
type of research, testing, and interdisciplinary consultants that may be
required, the initial hypotheses of the
cause of failure, and the anticipated

budget that will be required for fees and expenses.

The oral report may be given directly to the client's attorney, thus permitting the attorney to determine if the report contains negative findings with respect to the position of the client. The attorney can use this information to plan the case.

Some attorneys prefer that their experts prepare no written reports, and that they deliver all reports orally, in order to make their findings immune from a subpoena duces tecum. (See definition in Chapter 6.)

3.1.2 PROGRESS REPORTS

Interim progress reporting may be required by clients under the following circumstances:

3.1.2.1 Preliminary Report

An initial progress report or preliminary report may be required by a client who desires an overview of the progress of the investigation immediately after the initial site visit. The report should summarize the initial findings to date, including reporting of the initial site visit and the initial hypothesis of failure. It should also forecast the additional research, testing, and consultants that may be required. A statement discussing the preliminary nature of this initial report should be included. A budget for the anticipated scope of work should be developed and presented to the client in a separate letter. The material developed for the preliminary report should be used in the final report.

3.1.2.2 Interim Investigative Reports

Interim or progress reports may be required periodically during the failure investigation of large scale projects. The interim reports will update the progress of the investigation and provide a forecast of the scope of work to be accomplished prior to the next interim report.

3.1.2.3 Weekly Reports

During the course of a complex investigation involving the activities of testing laboratories and the activities of several interdisciplinary experts, the attorneys for the client may want weekly status reports to keep them informed of progress and of developments in the investigation.

3.1.2.4 Formats for Progress Reports

Progress reports, including preliminary, interim, and weekly reports, should be prepared using a format similar to that which will be used in the final report. Adopting this technique will facilitate the preparation of the final report by using the logical progression developed in earlier reports. A recommended format for reports is included in this Chapter. (Section 3.3)

3.1.2.5 The Final Report

The final report will contain a complete history of the investigation and its conclusions, including the research and analysis of the cause of failure, the findings of the investigation team, and the failure hypothesis developed. The report might also contain recommendations for further remedial action.

The final report will include all valid information included in early reports, either written or oral, in addition to the final summations and conclusions.

3.2 PREPARING THE WRITTEN REPORT

The written report will detail the salient components of the investigation. The report should be written using clear and concise writing. The thoughts of the forensic engineer should be carefully stated, using words and phrases that are not subject to misinterpretation.

The report should be written using a logical format that generally follows the same thought process used in the investigation, detailing salient events of the investigation from conception to completion.

All of the reports, including the final document, may use the following format:

1. Section One: Purpose:

    The purpose will include a summary of the salient aspects of the report, a statement of the overall purpose of the investigation, and identification of the client who sponsored the investigation.

2. Section Two: History

    The section on history will provide an overview of the history of the failure including the following information:

    (1) Description of the failure and chronological account of the investigation

    (2) Construction: date of initiation and completion of construction

    (3) Owner: name and address of the original owner and successors

    (4) Contractor: name and address of the general contractor

(5) Design Team: name and addresses of design team members, including architects, engineers, etc.

(6) Construction Documents: a complete listing of construction documents and other documents uncovered during the investigation and used in the investigation.

3. <u>Section Three: Description of Project</u>

This section will provide a general description of the project and a detailed description of the failed construction.

4. <u>Section Four: Observations</u>

A description of the site visits and a detailed description of site observations, field testing, sketches, photographs, and samples taken for laboratory testing are contained in this section.

5. <u>Section Five: Testing:</u>

An overview of the field and laboratory testing programs with testing objectives and a summary of the results included. Laboratory reports should be attached in the Appendix.

6. <u>Section Six: Analysis:</u>

An analysis of the failed construction and results of the synthesis of the data resulting from the investigation are included in this section. This section should include a description of the methodology used for the review, computations, evaluation, and development of hypotheses.

7. <u>Section Seven: Conclusions:</u>

This section will include a presentation of the most probable cause of the failure and discuss the relevant factors contributing to the cause.

8. Section Eight: Recommendations:

A statement of recommended action based on the requirements of the investigation includes:

(1) Remedial Action: If the client requested a direction for repair of the failure, the report will recommend a remedial program.

(2) Avoidance of Future Failure: Statement of how this failure investigation and its findings may be used to prevent future failures in this or other similar facilities.

9. Appendix

The Appendix contains detailed data utilized in the investigation report, including:

(1) Test Reports
(2) Contract documents (reduced)
(3) Field sketches
(4) Photos
(5) Author's biographical data

3.3 WRITING A CLEAR REPORT

In order to prepare a report that accurately reflects the findings of an investigation, the forensic engineer must develop a document that contains clear, concise writing and eliminates unnecessary technical jargon. This document will be read by your client and used to explain the cause of the failure to those to whom your client is accountable, and it may be used by the attorneys for the client in order to implement a lawsuit. However, the most important reason for using understandable writing is that this document might be reviewed by a jury of lay people during a legal contest. This jury will not have a technical background, and the members of the jury can understand the report only if it is written in understandable language. Thus, technical descrip-tions should be summarized in the body of the report using understandable language, with highly technical material detailed in the Appendix.

The preparation of an investigation report may offer an opportunity for the forensic engineer to explicate an exciting hypothesis developed during the investigation. However, to communicate his or her conclusions in an understandable manner, the forensic engineer must write clearly.

The engineering curricula in the majority of universities lack sufficient courses on literature and writing. The art of clear communication, however, can be developed with the aid of guidelines and reference materials.

A quality report depends on the ability of the forensic engineer to develop good writing skills. To assist the forensic investigator in developing these skills, this section includes guidelines, and recommendations for reference books that will aid in writing a comprehensive report.

3.3.1 AVOIDING THE USE OF ABSOLUTES

The thought process delineated in the report should be carefully worded in order to avoid misinterpretation. Avoid the use of absolutes (e.g., "never" and "always") unless they are required. Be positive when putting the message across to the reader.

3.3.2 USE OF WORDS

Words are the writer's tool, and the writer can learn to use them as a surgeon uses a scalpel: deftly and skillfully with an exact sense and a sure knowledge.

It is important to understand the difference between abstract and concrete words. An abstract word expresses an idea that can't be visualized. A concrete word (not to be confused with a building material) is one that causes an immediate picture to appear in your reader's mind. The majority of technical words are abstract and cannot be visualized by a layman. Nevertheless, attempts should be made to convert abstract technical terms to concrete terms. Imagine the effect certain words will have on your audience. Use concrete words so a specific picture will be created in the reader's mind. When a vivid and concise image is evoked by the writing, the reader will find the report interesting.

3.3.3 CLEAR COMMUNICATION

Putting words on paper to evoke specific images in the mind of the reader is the goal of the investigative writer; but the writer must control the images completely if the reader is to exactly understand the thought process. The forensic engineer as technical writer might assume that the reader will understand the majority of words contained in the report. After all, the readers and writers have grown up using the same language. This, however, is not quite true. Not everyone gets the same image from every word, no matter how clear the meaning of the word appears to the writer. This is especially true when technical language is used in a report that may be read by laypersons. Keep the writing in the report concrete and clear.

Keep a good dictionary and the thesaurus on hand while writing. If the investigative writer is the least bit uncertain about a word, then he or she should take a moment to look it up.

Using these basic guidelines, the technical writer will be able to express ideas more accurately, and the report will become more precise and understandable.

3.3.4 USE OF VOCABULARY

Another trap of the technical writer is to use a runaway vocabulary. If the reader is lost in a maze of words that can be deciphered only with a technical lexicon, then the writer is not communicating well. The written document is successful only when the reader knows what the document is about.

Read the final draft aloud, listen to your voice speaking the words as if they were dialogue. This technique will uncover un- necessary technical jargon and aid in drafting an understandable document.

3.3.5 WORD BOOKS FOR WRITERS

The technical writer needs at least two books to assist in the selection of words: a dictionary and a thesaurus. In addition, other guides are recommended for good writing.

The dictionary is used to build a specific vocabulary for writing. A thesaurus is a type of dictionary that contains synonyms rather than straight definitions. The thesaurus is most useful in tracking down a word that one can't quite remember. It is also of great assistance in finding concrete words to express abstract idea. The use of synonyms in report writing will make the work more readable and provide added confidence in the drafting of reports.

A great aid to the technical writer in enhancing communication and writing skills is a volume entitled, "The Elements of Style" by Strunk and White (New York MacMillan, 1965). This book contains a wealth of communication aids.

In addition, the following volumes are recommended:

"The Elements of Grammar" by Margret Shertzner
(New York, MacMillan, 1950);

"The Elements of Editing" by Arthur Plotnik
(New York, MacMillan 1982).

3.3.6 USE OF THE SENTENCE

The technical writer should attempt to develop skill in writing sentences that are clear and uncomplicated. It is more important to get ideas across than to demonstrate the writer's literary talents. Keep the sentences short and to the point.

3.3.7 USE OF THE PARAGRAPH

One purpose of a paragraph is to break up the type on a page. A full page of solid type can discourage the most courageous reader.

The paragraph can also be used to make the report more understandable, and can be used to separate individual ideas so they are easier to comprehend.

Strive to make the majority of paragraphs of moderate length. Some paragraphs may be a single sentence when the writer wishes to make an important point; but too many short paragraphs used consecutively will make the writing disjointed or unconnected.

3.3.8 REWRITING THE WORK

It has been said that good reports are not written, they are rewritten. There are few writers whose first draft is written well enough to be a final product without some adjustment. The majority of writers do a considerable amount of rewriting; often two or three drafts are required until the quality of the work is satisfactory. Do not be discouraged by drafting and redrafting the report until it clearly expresses your thoughts and concepts. Many successful writers rewrite several times until the final work is satisfactory.

Once the writer has completed the first draft of a section of the report, he or she should reread the work as if it is being read for the first time. The successful writer criticizes his or her work, rewriting it to improve its clarity and quality. When reviewing for the purpose of rewriting, be alert for excess words, improper structure, grammatical errors, and unclear phrasing.

3.3.9 TECHNIQUE OF EXPOSITION

Expository writing is the form used in the majority of textbooks, scientific papers, instruction manuals, and technical reports. Exposition is instruction or explanation.

It informs the reader. It increases the reader's knowledge and helps him or her understand the subject better. Exposition is the technical writer's method of conveying meaningfully a group of facts to the reader.

Expository paragraphs are written around a topic sentence. The topic sentence clearly states the main idea the writer wants to present in the paragraph. It is the key sentence to which all other sentences in the paragraph are related. It serves to put things into focus for the reader.

Since the topic sentence is the key to all expository writing, the most effective ones are strong, direct statements that can be understood with a minimum of effort. The writer may use more complex sentences to expand the topic sentence, but it is important to make sure the point of the topic sentence is understood.

The majority of expositional paragraphs begin with the topic sentence. This helps to achieve clarity and continuity. The second most important position for the topic sentence is at the end of the paragraph. When the topic sentence appears as a climax, the writer must build the beginning of the paragraph to meet it. After the introductory paragraph, the facts should be listed. The facts are more clearly understood when they are enumerated in a numbered list. The principal problem in the listing of facts is keeping the list from becoming dry and unpalatable. Try to find bright, concrete words that enhance the imagery of the text.

3.3.10 STRUCTURING THE REPORT: THE OUTLINE

Many technical writers complain that the most difficult part of writing is writing the first sentence of the first paragraph. But writing that first sentence first is not a recommended technique for developing a successful report. The first step is to create an outline. After the outline has been made, the writer should prepare summary paragraphs about the topics covered

in each section of the outline. The
paragraphs should then be rewritten until
they convey the information clearly and
concisely. The thought processes of the
engineer during the investigation should be
illustrated in the final report.

CHAPTER 4

INVESTIGATION OF GEOTECHNICAL FAILURES

This Chapter provides guidelines for preparing an investigation relating to Geotechnical Failures. The techniques and data in this section are presented in the general format outlined in Chapter 2.0. The general format is interfaced with specific methodologies for geotechnical investigations to assist the investigator in successfully developing an investigative plan that will identify appropriate failure hypotheses.

4.1 PLANNING THE GEOTECHNICAL INVESTIGATION

The implementation of the investigation of a geotechnical failure begins with the development of a plan that provides systematic examination of the failure, including the historical background of its design and construction as well as testing, synthesis of data, development of failure hypothesis, and preparation of reports. The planning and execution of the investigation will follow the guidelines for investigations described in Chapter 2.0.

This chapter may not reflect the complete state-of-the-art. The reader should be aware that there are other modes, causes, and methods of evaluation.

Following is an outline of some recommended procedures for the investigation of geotechnical failures.

4.1.1 PLANNING THE INVESTIGATION

The procedures for planning an investigation include (see also Chapter 2.0):
1. Client interface
2. Identification of the investigation team
3. Operations planning

4.1.2 SITE OBSERVATIONS AND ANALYSIS

The methodology for optimizing activities carried out during site visits to geotechnical failures are contained in this chapter (Section 4.2).

4.1.3   TESTING PROGRAMS

The process for identifying and developing testing programs, including field and laboratory testing, are contained in this chapter (Section 4.3).

4.1.4   DOCUMENT SEARCH

Refer to guidelines contained in chapter 2.0, Section 2.6.

4.1.5   LITERATURE SEARCH

Refer to guidelines contained in chapter 2.0, Section 2.8.

4.1.6   MODES OF GEOTECHNICAL FAILURES

Descriptions and examples of modes of geotechnical failures are outlined in this chapter (Section 4.4).

4.1.7   PROFILES OF GEOTECHNICAL FAILURES

Examples of profiles of geotechnical failures classified by type of super-structure and type of foundation are included in Section 4.5.

4.1.8   DESIGN ANALYSIS

Guidelines for reviewing the design of the distressed geotechnical structures are included in this chapter (Section 4.6).

4.1.9   INVESTIGATION SYNTHESIS

An overview of the methodology for synthesizing the information and data collected during a geotechnical investigation is contained in this chapter (Section 4.7).

4.1.10   DEVELOPMENT OF FAILURE HYPOTHESIS

Guidelines for development of a failure hypothesis are contained in this chapter (Section 4.8).

4.1.11 PREPARATION

Discussion of types of reports and techniques for report preparation are contained in Chapter 2.0.

4.1.12 LEGAL CONSIDERATIONS

The background and responsibilities of the forensic engineer engaged as an expert witness in a legal contest is contained in Chapter 6.0.

4.2 SITE OBSERVATIONS AND ANALYSIS

The goals of site visits are to make overall visual examinations and graphic and narrative records, to obtain eyewitness accounts, and to initiate the development of material testing and measuring programs.

As with all failures, distress related to subsurface construction requires the presence of a forensic engineer at the site immediately following the discovery of the event. Quick action in observing the failure condition will permit a firsthand review of the nature of the failure prior to any reconstruction or removal of debris. The initial observation will provide information about the nature of the failure and any aspects of it that require further study. This is the first step in determining causation and assigning responsibility.

Chapter 2.0 contains additional techniques for developing a body of site information about the failure.

4.2.1 PHOTOGRAPHS/VIDEO RECORDING

At the time of the first site visit, photographs should be taken by the investigator and/or a video camera should be used to depict all elements that may have led to the failure. Such pictorial representations should portray relationships of size. This documentation should be properly catalogued and indexed for subsequent review.

4.2.2 **NARRATIVE AND WRITTEN RECORDS**

> Extensive use should be made of notes
> and/or sketches, and portable dictation
> equipment to record what has been
> observed by the investigator.

4.2.3 **SAMPLES**

> Where available, samples of soil, rock,
> and water should be obtained, as well as
> specimens of any material that is
> believed to have contributed to the
> failure.

4.2.4 **EYEWITNESS INTERVIEWS**

> When possible, interviews should be
> conducted with those who may have
> observed the conditions prior to,
> during, or subsequent to the failure.
> See techniques for eyewitness interviews
> in Chapter 2.0.

4.3 **TESTING PROGRAMS**

Geotechnical investigations primarily refer to the
behavior of soil, as well as to rock and groundwater. Therefore, the testing in connection with
a geotechnical failure is principally directed to
field and laboratory tests on soil, rock, and
groundwater and their influence on underground
construction.

4.3.1 **FIELD TESTING**

> The testing in the field at the failure
> site is primarily accomplished by borings,
> test pits, observation wells, as well as
> soil and foundation load testing. The
> former is to determine the soil and rock
> profile and groundwater levels immediately
> adjacent to the failure location and the
> surrounding area. The latter is to eval-
> uate the load carrying capacity and
> settlement data of the soil at specific
> elevations and the load deformation charac-
> teristics of the foundations, particularly
> with reference to friction or end-bearing
> piles. The complete environment surrounding

-79-

the structure and adjacent area should be known and documented.

4.3.1.1   **Borings and Test Pits**

The subsurface investigation may be of the exploratory type, necessary for identifying the subsurface profile. Also borings may be necessary for obtaining undisturbed samples for future laboratory testing. The explorations should provide spoon samples for identification purposes; evaluation of soil compactness or stiffness through Standard Penetration Test resistance; soundness and quality of rock by coring; and levels of perched and static groundwater, measured by visual observation in the borings or by use of observation wells and/or piezometers. Borings can be used to obtain undisturbed samples of compressible soils and as yet undisclosed compressible soil strata. Excavation of test pits are usually helpful in evaluating shallow subsurface soil, rock, and groundwater conditions, and also for soil excavatability and rock rippability.

The types of borings in which samples (both spoon and undisturbed) can be obtained are:

1. cased borings
2. auger boring
3. rotary drilling
4. wash boring
5. percussion drilling

Applicable ASTM procedures should be used for obtaining undisturbed soil samples, performing Standard Penetration Tests, and the recovery of rock cores.

The Standard Penetration Test results are probably the most widely used method for determining a design basis. But advancing rig and tool technology coupled with changes in labor costs tend to create a less than standard test. Tool sizes and new tools vary from region to region. Frequently different contractors have different sizes and weights of tools. Many of these variations do not conform to the ASTM norm.

It has been argued that because of these variations, a great conservatism in design has resulted. It has been suggested that this conservatism or over-design needs reviewing. Perhaps the simple inexpensive Standard Penetration Test is something the engineering profession and the construction industry has outgrown. Potential variations in accuracy of data should be documented and expressed in all reports.

4.3.1.2   Seismic Exploration

Seismic methods for determining rock profiles or strata demarcations between dense and softer soils are relatively rapid and reliable.

4.3.1.3   Field Load Tests

Load tests can be performed on foundation elements that have undergone failure or on prototypes, installed at adjacent sites or in a laboratory, to duplicate the failed elements for a determination of ultimate load carrying capacity and deformation characteristics. Such field performance would include pile load tests, soil plate bearing tests, and load tests on actual

and representative spread footings. Load-deflection data would
indicate the ultimate and maximum
allowable foundation loading.

4.3.1.4   Dutch Cone Penetration Test

In this process, a cone is pushed
into the soil stratum and
resistance is measured. The
resistance may be of the cone and
the skin resistance of the pipe
segment that advances the cone.
The cone resistance is related to
the undrained shear strength of
the soil.

4.3.1.5   Field Density Tests

Such tests may be performed to
determine the in-situ soil
density at the failure site, as
compared to the laboratory
maximum density.

4.3.1.6   Test Pits

Evaluation of test pits can be
used to observe soil strati-
graphy, to obtain both disturbed
and undisturbed samples and to
observe groundwater levels.

4.3.1.7   Borehole Shear Test

In this test, a serrated cylinder
is split lengthwise and lowered
into an uncased borehole. The
cylinder is extended into the
opposite sides of the borehole by
pressure from the surface and the
lateral pressure is measured.
The pressure is incremented and
the test is repeated. When
sufficient data is obtained a
Mohr rupture envelope can be
plotted for shear strength
determination.

4.3.1.8   **Slope Indicators**

These instruments measure and record horizontal and vertical movement in soil and rock. From this information, properties of water and earth pressures and in-situ moduli of rock can be determined.

4.3.1.9   **Piezometers**

A piezometer is a device installed at varying depths to measure the pressure head of pore water at specific points within a soil mass. The transducers are usually made of inert plastics or porous stones so as to eliminate electrolytic reactions. Thus, data pertaining to slope stability and rates of consolidation of soft soils can be determined.

4.3.1.10   **Additional Testing**

Some additional field tests that might be conducted are:

1.   vane shear tests
2.   penetrometer tests
3.   tests fills, instrumented for settlement and pore pressure measurements
4.   slope stakes
5.   precise level surveys
6.   settlement observation by hose-level device
7.   tiltmeter
8.   fiber optic bore scope/bore hole camera
9.   field permeability tests
10. grout acceptance tests

4.3.2   **LABORATORY TESTS**

The testing of soil samples in the laboratory is for identification purposes and for obtaining strength and compressibility characteristics. The former includes such tests as moisture contents, grain-size distribution, plasticity of fine-grained

-83-

soils, and organic content. The latter includes consolidation tests for compressibility and unconfined and triaxial compression tests for strength, all performed on undisturbed soil samples. Permeability tests may also be conducted to determine seepage and drainage characteristics, and unconfined and high-capacity triaxial compression tests may be performed on rock cores for a determination of ultimate bearing capacity. Chemical tests may be made on groundwater and soil samples to evaluate their chemical composition. Compaction tests may be performed to obtain maximum density and optimum moisture data. All testing should be in conformance with such applicable standards as ASTM, AASHO, and others.

If special tests are required, these may include:

1. Swell
2. pH level
3. Sulphate or other salt contents of soil
4. Electrical resistivity
5. Chloride level and corrosion potential

4.4   MODES OF GEOTECHNICAL FAILURES

The following are some primary modes of geotechnical failures that may be considered by the forensic engineer in the course of the investigation. However, there may be other modes of geotechnical failure that are not included in the following list.

4.4.1   SETTLEMENT

Settlement is vertical or differential movement of the failed facility resulting in distress or collapse. Evaluation can be made through field and laboratory testing (particularly consolidation tests) and by measurement and comparison to actual applied loading and/or the existing nondistressed portion of the structure.

4.4.2 STABILITY FAILURES

Surficial or massive earth movements resulting in embankment slides or collapse of trench walls are classified as stability failures. Data for evaluation can be obtained primarily through unconfined and triaxial compression laboratory tests for a determination of shear strength in conjunction with embankment slopes and heights, and applied surface loading.

Support of foundations on different materials can cause uneven settlement, rotation and/or tipping (e.g, foundation bearing partially on sound rock and partially on a softer material, or structures partially pile supported). In non-uniform settlement due to removal of foundation elements and/or excavating below or adjacent to structures, both lateral and vertical support can be compromised.

4.4.3 BEARING CAPACITY FAILURES

Vertical and differential movements caused by exceeding the ultimate shear strength of the soil beneath the failed foundation are classified as bearing capacity failures. Laboratory testing for soil shear strength (unconfined and triaxial compression tests) and determination of applied foundation loading can be the basis for evaluation.

4.4.4 LATERAL EARTH PRESSURE FAILURES

Horizontal movement, tilting, or collapse of retaining structures due to excessive lateral earth and hydrostatic pressures may result from exceeding the soil shear strength behind or below the retaining facility. Evaluation data can be obtained from laboratory shear strength and consolidation tests, data on water level, and groundwater level fluctuations beside or beneath the structure, combined with knowledge of applied surface loading. This failure may also be influenced by a structural failure of components of the retaining facility.

4.4.5   **FAILURES DUE TO EXCESSIVE SOIL YIELDING**

Vertical and horizontal movements resulting in an inoperative facility or one in severe distress may be due to excessive soil deformations without an actual collapse. Data for such an occurrence can be obtained from field and laboratory tests and by measurement and examination of the applied loading.

4.4.6   **FAILURES DUE TO UNANTICIPATED STATIC AND DYNAMIC LOADING OR UNKNOWN CONDITIONS**

The former would include facilities that may have been subjected to loading that was unanticipated during design and resulted in failure or distress. Typical examples of the latter would be uncharted underground mines, cavities, pipes, tunnels, or other underground structures.

4.4.7   **UPLIFT**

Uplift caused by hydrostatic pressures must be taken into account and determinations must be made as to groundwater levels and differential hydrostatic pressures.

4.4.8   **DRAGDOWN**

Piles, fully or partially deriving their support by friction in compressible soils that can consolidate due to surface loading or dewatering, may develop future dragdown forces. These forces, if not accounted for in design, may cause eventual pile settlement.

4.4.9   **FAILURE DUE TO WATER FLOW**

Control of water is important when such flow occurs under footings or foundation walls. Otherwise, the footings or foundations, undermined by erosion, could cause failure and washouts.

4.4.10   **FAILURE DUE TO SHOCK AND FOUNDATIONS**

Vibrations on foundations supporting engines, compressors, punch presses, generators, etc. may cause them to undergo

settlements larger than those imposed by
static loads. Vibrations can affect the
entire building system, including plumbing
and personnel comfort, unless controlled.
Internal friction and other factors should
be taken into account in foundation analysis and design.

4.4.11 STRUCTURES FOUNDED ON METALLIC SLAGS

Open-hearth slag is an inexpensive fill
material found in areas that are near metal
manufacturing plants. Permeability and
absorption characteristics of slag may
cause differential settlement and potential
expansion in structures and heaving in
parking lot pavements.

4.5 PROFILES OF GEOTECHNICAL FAILURES

To assist the forensic engineer in identifying
potential causes of failures, the following is
provided as an overview of classic geotechnical
failures categorized by the type of facility and
the type of structure.

The geotechnical failure modes described in the
preceding section may be applicable to all aspects
of underground construction and are not limited
only to structural foundations. The following is
a partial listing of the facility types that may
experience failures.

4.5.1 BUILDING AND BRIDGE FOUNDATIONS

Examples of such failures may be found in
the following foundation types.

4.5.1.1 Spread Footings

Plain or reinforced concrete pads for
the support of building walls and
columns and for bridge piers and
abutments.

4.5.1.2 Mat Foundations

Where building columns and walls are
supported on large rigid or flexible
reinforced concrete pads that

distributes the loading to the underlying soil.

4.5.1.3 **Eccentricities**

Eccentricities not considered in design but built into the project during the construction and post-construction phase can cause differential settlement in the foundation and overloading of the super-structure system.

4.5.1.4 **Pile Foundations**

For column, wall, pier and bridge abutment support, sometimes the structural loading is transferred through friction piles and/or end-bearing piles to soil or rock. Failures may be found for the following pile types:

1. **Steel Piles**: These pile types include H-bearing and closed-end concrete filled and open end pipe.

2. **Timber Piles**: Included are creosoted and non-creosoted wood of varying tip diameters.

3. **Concrete Piles**: Such piles could be cast-in-place or precast concrete or composite (concrete and timber) members.

4. **Prestressed Concrete Piles**: For piles requiring high carrying capacity, prestressed concrete may be used.

5. **Pressure Injected Piles**: Concrete may be dynamically installed to densify granular soil at a depth below ground surface and to provide support through soil-concrete interaction.

6. Caissons:

- Open-End caissons: May be any crossectional shape to provide resistance to vertical loads and lateral pressures.

- Closed-end or box caissons: Differs from open-end by incorporating a base. Frequently these are large enough to be used alone to support a large concentrated load and they can be cast to eliminate footings and pedestals.

- Pneumatic caissons: Airtight enclosures that employ air pressure to maintain a cavity in the excavation area.

- Drilled caissons: Advanced by a cylindrical cutting tool and cased with perimeter forms and rings. These can be used when vibrations and displacements from pile driving are critical. Uplift resistance and increased capacity can be provided by belling or undercutting. These foundations usually carry larger loads than piles and are easier to inspect.

4.5.1.5 **Floor Slabs on Grade**

While such reinforced and non-reinforced concrete slabs do not usually carry column and exterior wall loading, they are subject to dead and live point and area loads and partition loading.

-89-

4.5.1.6   Miscellaneous Considerations

At times, foundation failures are attributable to the following occurrences.

- Adjacent Construction: The influence on buildings and bridge foundations of excavation, foundation construction, and dewatering for adjacent facilities may affect an existing project.

- Concrete and Steel Deterioration: Due to chemical and other influence on concrete, steel reinforcement, and steel piles and caissons may deteriorate prematurely. These influences may include polluted ground and water areas and sanitary landfills.

- Frost: The effect of sub-freezing weather on concrete and the heaving (and eventual subsidence) of soil beneath pavements and foundations may be a factor.

- Expansive Soils: The swelling and subsidence of soils due to fluctuating groundwater levels.

- Buckling: Piles or sheet piling under load which may or may not be completely embedded, or sufficiently supported laterally by the soil.

4.5.2   EARTH EMBANKMENTS

Failures resulting from surficial and massive earth and mud slides.

4.5.3   RIGID AND FLEXIBLE PAVEMENTS

Failures in such facilities may occur due to the following causes:

4.5.3.1 Subgrade Soils

Failures may result from loose, soft, expansive or frozen subgrades.

4.5.3.2 Materials

Failures may be influenced by improper soils or compaction used in the subbase and base course, and inadequate materials for pavement and wearing courses.

4.5.3.3 Thickness

Insufficient thickness of one or more elements in the pavement section may result in a failure.

4.5.3.4 Groundwater

Groundwater fluctuations can influence pavement behavior as well as frost action beneath the pavement and the swelling of expansive soils.

4.5.3.5 Loading

Overload or unanticipated loading on the pavement may cause distress.

4.5.4 RETAINING STRUCTURES

Failures in retaining walls and crib walls may be due to the following:

4.5.4.1 Foundation Soil Failure

Settlement of the base or toe.

4.5.4.2 Backfill

Improper choice of material or inadequate compaction.

4.5.4.3 Overload

Excessive loads on the backfill surface or loads applied in too close proximity to the retaining structure.

4.5.4.4   Structural Design

Inadequate design of the base or wall.

4.5.4.5   Hydrostatic Forces

Excessive water pressures behind the retaining structures and inadequate drainage openings through the wall.

4.5.4.6   Vibratory Forces

The proximity of machinery and other vibrating elements.

4.5.5   BULKHEADS AND COFFERDAMS

While the failure characteristics of bulkheads and cofferdams are similar to retaining structures, several additional factors may cause failures.

4.5.5.1   Tie-Backs and Anchorage

Insufficient or improper anchorage systems and inadequate size and spacing of tie-backs.

4.5.5.2   Material

Deterioration of timber, steel, concrete, or hardware and improper strength of piles, walers, sheathing, and tie-backs.

4.5.5.3   Installation

Construction techniques as influencing behavior in service.

4.5.5.4   Blow-Outs

Hydrostatic uplift may occur during cofferdam construction.

4.5.6   TUNNELS

Failures in this type of construction, either by open-cut or by boring techniques, may result from the following:

4.5.6.1 Earth Pressures

Vertical and lateral soil pressures may be greater than anticipated and unanticipated pockets of soil or voids may be encountered.

4.5.6.2 Floor or Pavement

Adverse soil conditions or uplift hydrostatic pressures may influence behavior of vehicular or railroad tunnels.

4.5.6.3 Unanticipated Rock Type

This can cause difficulties during horizontal boring and in creating varying pressures on the tunnel.

4.5.6.4 Water and Gas Seepage

Water and gas seepage during tunnel construction can result in hazardous and difficult control problems.

4.5.7 UNDERGROUND PIPING

Many failures occur during and subsequent to the installation of storm water and sanitary sewers. Some of the causes are listed below:

4.5.7.1 Pipe Support Failure

Inadequate bedding or support of sewer lines can lead to pipe subsidence and deflection.

4.5.7.2 Groundwater

This would include difficulties in dewatering during construction and softening of the foundation and side-wall soils due to soil/water interaction and change in soil characteristics.

4.5.7.3   Construction Techniques

Eventual failures may result from using improper construction procedures during the installation of the pipe, sheeting, manholes, and other appurtenances.

4.5.8   TRENCH EXCAVATIONS

Failures commonly occur during construction in excavating trenches for sewer, foundation, and other underground construction. Some of the causes are as follows:

4.5.8.1   Depth of Trench

Unsupported height of trench may be excessive and result in a cave-in during construction.

4.5.8.2   Sheeting

Where sheeting is used in the excavation, it may be of insufficient depth or strength or may not be properly braced.

4.5.8.3   Dewatering

Groundwater may be difficult to control and can weaken subgrade and embankment soils.

4.5.9   GROUNDWATER VARIATIONS

The effects of seasonal groundwater variation and high intensity rainfalls can influence all types of underground construction. Dewatering to control such water fluctuations during construction may adversely influence adjacent foundations.

4.5.10   SEISMIC EFFECTS

Mass earth movements, including earthquakes, can cause excessive distress and possible failure in underground construction if not properly considered in design.

4.5.11 **COMPACTED BACKFILL**

Foundations are frequently constructed on engineered controlled backfill. Failures may occur due to the following causes:

4.5.11.1 **Fill Subgrade**

If all unsuitable material is not excavated, subsidence could develop after the installation of the compacted fill and the superimposed loading.

4.5.11.2 **Backfill Material**

Improper or non-uniformly graded backfill soil either may be difficult to compact or may contain objectional material that could be subject to eventual subsidence.

4.5.11.3 **Compaction**

If adequate compaction is not achieved to obtain the desired soil density, the backfill can cause difficulties in the installation and future settlement of the backfill material.

4.5.12 **ENVIRONMENTAL CONCERNS**

Pollution of soil and groundwater has caused many environmental problems and has affected not only underground construction materials, but has had a detrimental influence on potable water supply and spread of disease.

4.6 **DESIGN ANALYSIS**

Independent analysis of the design of the original facility should be carried out to ascertain the ability of the distressed facility to satisfy building codes and service requirements. This design review should identify incidents of errors or omissions in the design documents. If records indicate that the facility has been modified from its original design or that the service loading had to be altered from that of the original design, then these factors must be considered in the design analysis.

See Chapter 2.0 for additional guidelines relative to design analysis.

4.7 INVESTIGATION SYNTHESIS

As data and observations are collected from the efforts relating to field observations, document search, literature search, testing programs and design analysis, the process of synthesizing this body of information into a failure hypothesis should be implemented.

Following is a listing of specific appraisals of data that will lead to the efficient development of failure hypotheses. Each listing will include reference to other sections that will guide the investigator.

4.7.1 DEVELOPMENT OF A HISTORY OF THE DESIGN AND CONSTRUCTION OF THE FACILITY

Use material collected in the document search (Chapter 2.0) to provide an overall narrative history. Refer to Chapter 3.0, (The Report) to review historical data required for the report.

4.7.2 DESCRIPTION OF SITE AND SERVICE CONDITIONS AT THE TIME OF FAILURE

Use materials collected in the Document Search (Chapter 2.0), Eyewitness Accounts (Chapter 2.0), and Field Observations (Chapter 4.0) to develop a narrative account of the salient conditions prevailing prior to the failure.

4.7.3 DESCRIPTION OF THE FAILURE

Use Eyewitness Accounts (Chapter 2.0), Field Observations (Chapters 2.0 and 4.0), and Test Program Results (Chapter 4.0) to describe the sequence and mode of the failure.

4.7.4 DESIGN REVIEW

The development of a review of the design of the structure and foundation system as originally designed and/or modified will aid in the comprehensive overview of the failure. Use the guidelines contained in Chapter 2.0 to assist in developing a comprehensive design review.

4.7.5 DEVELOPMENT OF A FAILURE PROFILE

The development of a narrative and graphic overview of the failure profile can be carried out using the results of the testing program (Section 4.3), modes of geotechnical failures (Section 4.4), profiles of geotechnical failures (Section 4.5), and design analysis (Section 4.6).

4.7.6 CATEGORIZATION OF THE FAILURE

Using the analysis collected in the synthesis, the failure can be classified based on the information included in Section 2.0, The Investigation.

4.8 DEVELOPMENT OF THE FAILURE HYPOTHESIS

The culmination of the investigation will be the development of a viable failure hypothesis leading to a statement of the most probable cause or causes of the failure. The investigator should use Chapter 2.0 to aid in the development of failure hypotheses.

4.9 THE REPORT

The preparation of a clear, concise report is essential to relate the history of the investigation. Chapter 3.0 (The Report) will aid in the preparation of the report.

CHAPTER 5

INVESTIGATION OF STRUCTURAL FAILURES

This chapter provides guidelines for preparing an investigative program relating to the failure of engineered structures. The material in this section is presented in sequential order to assist the forensic investigator in successfully developing a logical process for establishing failure hypotheses during the investigation. The information contained in this chapter uses the general methodology outlined for the overall investigation contained in Chapter 2.0.

The guidelines for investigation of structural failures progress from the development of the investigative plan and initial site visit through field and laboratory testing to typical failure profiles classified by project types, structural type, connection type, and material type. Furthermore, the guidelines include methodology for synthesizing the data and developing failure hypotheses.

5.1   PLANNING THE INVESTIGATION

The investigation into the cause of a structural failure consists of a systematic examination of the failure, the historical background of the design and construction of the facility, its service life, and environmental effects related to the failure.

The planning and execution of the investigation will follow the guidelines described in Chapter 2.0. Following is an outline of the general procedures for the investigation of structural failures.

5.1.1 PLANNING THE INVESTIGATION

Guidelines for planning the investigation (see also Section 2.1) generally include:

1. Client interface
2. Identification of the Investigative Team
3. Operations planning

5.1.2 DOCUMENT SEARCH

Refer to Section 5.4 and Chapter 2.0 for guidelines relative to the document search.

5.1.3 LITERATURE SEARCH

Refer to Section 5.5 and Chapter 2.0 for guidelines relative to literature search.

5.1.4 REVIEW OF THE STRUCTURAL DESIGN

An overview of the rationale and techniques for reviewing the structural design of the original construction and modified construction is described in Section 5.6.

5.1.5 FAILURE PROFILES

To assist the forensic engineer in identifying potential causes of failure, Section 5.7 provides an overview of classic structural failure modes. The failure profiles are classified by project type, structural system, type of material, and type of connection.

5.1.6 INVESTIGATION SYNTHESIS

The overview of techniques for developing an investigation synthesis is contained in Section 5.8.

5.1.7 DEVELOPMENT OF HYPOTHESIS

Guidelines are contained in Section 5.9.

5.1.8 PREPARATION OF REPORT

The techniques for preparation of a clear concise report are contained in Chapter 3.0.

5.2 SITE OBSERVATIONS AND ANALYSIS

The forensic engineer should visit the site to personally ascertain the conditions pertaining to the failure or to performance problems. The goals of the site visits are to conduct overall visual examinations, to collect graphic and narrative records, eye-witness accounts and to develop test programs.

Following are typical techniques of conducting site observations and acquiring related data.

5.2.1 INITIAL SITE VISIT

The initial site visit should be carried out in order to evaluate the scope and nature of the failure, and to prepare the investigative plan. Please refer to Section 2.5 for methodology with regard to the initial site visit. This should include information on the following:

1. equipment for site visits
2. protocol for arrival at the site
3. safety and rescue operations
4. removal of debris
5. orientation of debris

5.2.2 DOCUMENTATION OF VISUAL EXAMINATION

The compilation of graphic and narrative records by the investigative team should be implemented to provide an in-depth overview of the failure scenario.

Section 2.5.8 describes in detail techniques of data collection that should be utilized during site observations, including:

1. sketches
2. reference systems
3. scale and orientation
4. types of sketches
5. photographs
6. videotaping
7. verbal descriptions
8. eyewitness interviews

5.2.3 EXPANDING THE SITE INVESTIGATIVE PLAN

To expand the data available at the failure site, in addition to narrative and graphic data generated by visual observations, the principal investigator should develop a detailed investigative plan that will implement additional types of data acquisition, including:

1. testing of Materials: field and laboratory testing
2. load testing of structure
3. collection of field documents

The decision to use any or all of these means of data acquisition is the responsibility of the principal investigator. Several other aspects should be considered during the entire testing program in order to develop a comprehensive and legally defendable body of information:

1. Most tests of significant nature should be observed by witnesses representing various interested parties and competent to judge the testing.
2. Reliance on more than one test method for especially sensitive tests.
3. No single discipline should be allowed to control major portions of the test program.

The testing program may include the following:

1. physical and chemical tests on materials
2. structural load tests on elements or element connections
3. structural load tests on major system components

The test program goals may include the gathering of data relating to several hypotheses dealing with material quality and defects, workmanship, unstable materials, and exposure to deleterious substances. By performing physical and chemical tests on a significant number of samples removed from the project, one can ascertain if in-situ materials were deficient. The principal investigator should specify individual tests by ASTM designation when possible, and request that the laboratory test the material in question for any other physical or chemical properties that could adversely affect its performance in the structure.

5.2.4 OVERALL VISUAL EXAMINATIONS

The initial visual examination of a failed structure will take place upon arrival at the site. It should be performed, no matter how briefly, by the principal investigator prior to preparation of the first organization and planning meeting with the team at the site.

During the initial visual examination, the principal investigator should expand and revise his or her original perception of the structure description and failure profile characteristics. The initial description of the collapse, the nature of the collapse, the nature of the structure, and the extent of damage may have been conveyed by the client's representatives who usually are not structural experts. Therefore, the initial description of the failure may be misleading or incomplete.

The principal investigator should prepare an overview description of the structure and the failure. He or she should prepare narrative and graphic observations that describe salient features of the debris. A great aid to the investigation would be for the principal investigator and the team to recreate sketches or photographs of the structure in its undamaged geometry, i.e., to relate the observed debris to the spatial orientation of the undamaged structure.

If possible, the live loading and direction of the load at the time of the failure should be assessed for use in the analysis.

Initial observations, both narrative and graphic, should include comments pertaining to the damaged and undamaged structure, as follows:

1. changes in plan location of structural systems or elements
2. change in vertical location of structural systems
3. structural members or connectors that are missing or distorted
4. changes in geometrical alignment of structural members

5. irregular or altered spacing of structural members
6. changes in curvature and unusual
   deformations of structural members
7. changes in the color of structural
   elements
8. changes in material conditions

During the initial site visit these
observations and graphics might appear to be
unrelated to the failure, but later reveal
illuminating and productive insights into
the cause of the failure.

5.2.5 <u>SITE CONDITIONS, SERVICE LOADING
OR ENVIRONMENTAL LOADING AT FAILURE</u>

The review of the loading of the structure
at failure can lead to evidence directly
contributing to the distress. Following are
several methods of establishing the load
acting on a structure at failure:

1. recording the vehicle's contents and
   material quantities entangled in the
   fallen debris and later calculating the
   respective loads

2. records of rain, snow, ice, wind
   velocity, and gust speeds from the
   National Climatological Survey or some
   other source (often wind velocities can
   vary significantly within short distances
   and over short periods of time)

3. past traffic volume records (for estimating repeated load cycles)

4. eyewitness interviews of people observing
   other activities around or under
   structure, but not directly on it

5. records of material or equipment stored
   at time of collapse

6. records of material hazards such as flood
   depth data, seismograph data, etc.

The objective of establishing the loading
condition at failure is to determine if

-103-

overloading was a factor in the failure or collapse. Even where a condition other than load is responsible for diminishing the strength or stability of a structural component or system, generally some increase in load may contribute to the collapse. It is very important to be able to define the level of loads just prior to collapse.

5.2.5.1   **Position of the Structure**

The overall movement and disposition of the structure during failure reflects the nature of the distressed condition. Documentation and evaluation of the geometrical configuration of the structure should include the following:

1. fallen portion including its size, orientation, and condition (intact or fractured)

2. locations and conditions where each end or section of the failed portion separated from the portion still standing

3. discontinuities that may be related to the separation at that point, such as: expansion joints, cold joints, foundation system variation, etc.

4. variations from assumed constructed position: for example, deflections and displacements from the assumed original position

5. movement related to adjacent structures, such as settlement, cracking, or tilting

6. variations between the configuration of the fallen or failed portion and the standing portion

7. variations in apparent loading at failure between the fallen and standing portion

8. material conditions
   (degradation or corrosion)

5.2.5.2 **Distress of Individual Structural Members**

Evaluation of information pertaining to individual elements that appear to be distressed includes the following:

1. deflection or deformation of the member
2. distortion of the member cross-section
3. dislocation and damage at connections
5. fracture surfaces
6. discoloration of surfaces such as water or rust stains
7. flaking or cracking of surface coatings such as paint

Variation of typical patterns are of particular interest. Atypical behavior or distress is most often determined by comparing and contrasting distressed elements with others in the structure and with other similar elements in the investigator's experience.

5.2.5.3 **Fracture of Surfaces**

The examination of fracture surfaces can yield valuable information regarding material performance during and before failure. All materials will typically respond to force systems in a predetermined reliable fashion. For example, failure surfaces of quality concrete, when failing in tension, will crack so that the failure plane runs through the sand-cement matrix and through the large aggregate. If the failure surface circumvents the aggregate, various cement paste deficiencies are likely. Carbon steels subjected to static tension show both a necking-down near the transgranular fracture surfaces. Often fatigued surfaces have lines radiating outward from a flaw. Intergranular fracture modes are more prevalent with corrosion and stress-corrosion.

Fracture surfaces are best studied by specialists. The investigator should be aware of the importance of fracture surfaces and, where possible, find all mating pieces of important fractures so that they can be studied by specialists. The investigator should be equipped with materials to protect fracture surfaces. For example, a simple clear lacquer may be used on steel fracture surfaces to prevent corrosion and degredation of the surfaces during storage.

5.2.5.4   **Condition of Materials of Construction**

The surface condition of materials often reflects the quality of the materials. The following are examples of material degradation:

1. thick layers of rust and corrosion of steel members
2. cracks with reddish-brown staining inside and beside the cracks in reinforced concrete
3. timber stained to a brownish or whitish color
4. spalling or chipping of edges of brick or concrete masonry at large weathered cracks or where there is a displacement in the plane of the surface

It is important to note that varying degrees of material degradation can be found on most older structures, but the investigator must determine the conditions that would significantly lower the structure's capacity or be directly related to the failure.

5.2.5.5   Evidence of Degradation

The effects of environmental conditions in which a structure exists can
affect the strength of a structure
and reduce its life. Following is a
listing of facilities and geographic
locations that can be exposed to
severe environmental conditions:

1. chemical plants
2. sludge treatment plants
3. bridge decks in northern latitudes
4. facilities exposed to salt water
   (both warm and cold climates)
5. food processing plants
6. other material processing and
   storage plants
7. parking garages

5.3   DEVELOPMENT OF TESTING PROGRAMS

Attention should be given to the planning of the
testing program required for the investigation,
including structural load tests, model testing,
field testing, and the acquisition of samples for
laboratory testing. Testing laboratories familiar
with sampling and testing according to ASTM
standards for the particular material in question
should be engaged as part of the team to carry out
sampling and testing. Consultation with these
firms will aid the investigator in developing the
total scope of testing required to carry out a
comprehensive test program.

Following is a summary of the applications of each
type of testing followed by a discussion of the
test's salient features.

5.3.1 TYPES OF TESTING FOR STRUCTURAL FAILURES

5.3.1.1 Field Testing of Materials of
         Construction

In the investigation of structural
failures the field testing of
materials is used to provide specific characteristics of the materials
of construction while in place.

Field testing of materials may be conducted on metals, concrete and timber.

5.3.1.2 **Laboratory Testing of Materials of Construction**

Testing materials in the laboratory enables a wide range of operations to be conducted with much greater efficiency and accuracy than if they were performed in the field.

Samples of materials are collected in the field and tested using the controlled conditions of a laboratory.

5.3.1.3 **Structural Load Testing**

Structural load tests are performed to assess the load-carrying and deformation characteristics of a structural element or complete structural systems. Structural load tests may be carried out at the site or in a laboratory.

5.3.1.4 **Model Testing**

Scaled-down models for structural testing can be used to simulate a complete structure or a structural component. Established rules of similitude must be carefully followed to achieve accurate quantitative results.

5.3.2 **MATERIAL TESTING**

5.3.2.1 **Field and Laboratory Testing**

The test program goals may include the gathering of data about material quality and defects, workmanship, unstable materials, and exposure to deleterious substances. Performing physical and chemical tests on samples removed from the project site can help evaluate the ability of in-situ materials to meet project

specification standards or code
requirements.

Laboratory reports should be
explicit in describing the tests
performed, results obtained, and
interpretations made with regard to
the significance of the findings on
the performance of the materials.

Following is a discussion of the
types of tests on materials commonly
performed during investigation of
structural failures. A variety of
testing may be performed in the
field on in-situ material. However,
a comprehensive testing program
should usually be performed in the
controlled condition of a testing
laboratory.

5.3.2 FIELD TESTING

Following is a discussion of tests that may
be conducted in the field to assess the
properties of in-situ materials:

5.3.2.1 Tests of Metal Components

The testing of metal components can
be achieved through the following
techniques:

1. Ultrasonic Testing: Ultrasonic
   detection methods use the
   transmission of ultrasonic pulses
   through the metal to detect the
   location of flaws. This technique can uncover defects in
   metal components or in welded
   connections.

2. Magnetic Testing: Magnetic crack
   detection uses the disturbance in
   a magnetic field to define crack
   locations and their dimensions in
   the base metal or in welds.

3. Radiation Testing: Radiation
   techniques are sometimes used to
   produce a film that would detect
   flaws in welds or in the parent

material. This technique may produce a three-dimensional view.

4. Hardness Testing: Hardness tests provide results that may be correlated with yield strengths. Test results of this nature may not reflect yield strengths of the interior of rolled shapes, but will indicate the strength of the surface material. Coupons may be cut to obtain yield strengths representative of the overall material.

5.3.2.2 Tests of Timber Components

The most common testing of timber involves boring into the surface with power hand drills or core drills for sample removal. The rate of penetration of the drilling operation may indicate low density or voids resulting from fungi, insect, or bacterial action.

5.3.2.3 Tests of Concrete Components

Field testing of concrete may include the use of a Swiss hammer to indicate the strength of the concrete. This device will provide approximate strength of the concrete, which should be verified with cores tested for compressive strength in the laboratory.

5.3.2.4 Tests of Masonry Components

Masonry materials testing may involve masonry units of clay, stone or concrete, mortar and the auxiliary steel, sealants, flashing, and coatings. Individual specimens of these materials can be tested in the laboratory, or masonry assemblages can be tested either in the laboratory or in-situ.

5.3.3 LABORATORY TESTING OF MATERIALS

The laboratory testing of materials offers a controlled condition to carry out a wide range of tests on materials of construction. The program for laboratory testing of materials begins with the systematic removal and collection of material samples at the site and extends through the testing procedures at the laboratory and preparation of test reports.

5.3.3.1 Field Sample Removal for Laboratory Testing

The collection of sufficient, representative, unbiased, random samples of construction materials for use in the laboratory testing program will follow the investigation plan established by the forensic engineer. The testing laboratory retained as part of the team will provide recommendations for collecting the field samples that will be transported to the laboratory.

5.3.3.1.1 Types of Materials

Some common methods of sample removal from structures, and the characteristics and properties that can be determined from the tests are as follows:

1. Metal Samples

Metal samples are obtained by coring, torch cutting, or sawing with a diamond wheel. Cores from metal components may be used to assess the strength and metallurgical composition of the material.

2. Concrete Samples

Concrete cores are used to assess chemical properties, strength of concrete, splitting tensile strength, the presence of delamination and as a sample source for petrographic studies. Core diameters usually vary in size from 2" to 8" or more. Concrete saws may be used to remove large samples. These samples may be used to examine the representative condition of the concrete.

3. Timber Samples

Samples from timber structural members may be taken by collecting cores. Cores can be used to evaluate the strength of the wood and can be examined for the presence of organisms.

4. Masonry Samples

Masonry samples taken from the structure might include individual masonry materials or masonry assemblages. It might also be necessary to obtain new specimens of the cement or sand that was commonly used in the mortar at time of construction to assist in analysis of hardened mortar taken from the building to determine the proportions of cement, lime, and sand used.

5.3.3.1.2 Labeling of Samples

It is necessary that a uniform system of numbering and handling samples be developed. It is advisable to categorize the samples generally, according to the type of testing to be performed, or for preparation for testing. Samples should be entered in a log that assigns a category number, location, type of sample (core, coupon, size), sample conditions on removal, type of testing required, photograph number, data acquired and by whom. Once the sample is withdrawn and logged, it should be marked or labeled using a label that adheres to the sample. All samples should be properly packed and stored to protect them from damage. The importance of a clear, concise and efficient nomenclature is critical for the orderly storage and easy retrieval of the samples.This will aid in the preparation of the final report.

5.3.3.2 Types of Laboratory Tests

Following is a listing of frequently employed tests for various materials of construction. The tests in each case relate directly to the capability of that material to perform according to industry standards.

5.3.3.2.1 Metal Testing

To ascertain specific characteristics of metal, the following tests may be used. It is important to select a testing laboratory that has facilities to properly carry out and assess the appropriate testing.

1. Charpy V-notch test to detect a tendency toward brittle fracture

2. metallurgical examinations to determine the percentage of critical elements that will verify conformance to standard specifications

3. chemical Tests to determine constituents detrimental to performance such as cracks, flaws, and occlusions

4. microscopic tests

5. electron microscope

6. X-rays

5.3.3.2.2 Concrete Testing

The following tests may be performed on concrete:

1. compressive strength tests using standard cylinders or cores

2. long term creep tests under controlled humidity and temperature

3. long term shrinkage/expansion tests under controlled humidity and temperature

4. air content tests

5. aggregate matrix microcracking

-114-

6. moisture control tests

7. photographic studies

8. split tensile strength tests

9. modulus of elasticity tests

5.3.3.2.3 Wood Tests

Tests may be carried out to ascertain the following data:

1. organic chemistry involved with timber products

2. interaction of timber treatment chemicals, fungi, and insects

5.3.3.2.4 Masonry Tests

Tests on masonry may include prisms in compression, flexure, or shear. Masonry units may be tested for compression, water absorption, freeze-thaw resistance, and expansion due to temperature changes, moisture absorption, or freezing. Mortar may be examined for cement and air content.

5.3.4 STRUCTURAL LOAD TESTS

Structural load tests may be performed to verify the load-carrying capability of a structural system or a structural element. In failure investigations where a portion of a structure has failed, field load tests may be performed on undamaged similar structural systems, or structural elements at the site. Depending on the type of structure under study the test may consist of static or dynamic load testing (impact or vibration). These tests may be considered necessary to determine the distribution of forces within a structure, the stress strain levels and stress ranges at design or at actual service levels, and the deflections and other displacements or rotation when analytical methods alone cannot accurately reflect the actual structural response. This may be the case when:

1. Severe deterioration has reduced or eliminated primary load paths. The existence of a secondary load transfer system is likely.

2. The uniqueness of analytical techniques necessitates verification by load test.

3. Construction workmanship or materials are different from those specified, and variability of materials prevents analysis from representing the analytical prototype.

The methodology for conducting structural load tests can be developed by the testing laboratory. The nature of the loading systems used in the tests must reflect the nature of the loads superimposed on the structure. In any case, a structural analysis should both precede and follow any load tests.

In failure investigations, laboratory load tests may be performed on structural elements removed from the structural system which failed. When performing load tests in the laboratory the investigator must be fully cognizant of variables, either excluded or altered when considering laboratory tests as compared to field tests. These factors include restraints from foundations, substructures, or other parts of the superstructure, volume changes, shrinkage, creep or relaxation of materials, ambient and intra-material moisture changes, and other geological or climatological conditions.

5.3.3.4 Model Tests

Model tests can simulate an entire structure or significant portion in a scaled-down size, so that:

1. The existing structure is not subjected to a test load which may render it useless or unrepairable after testing.

2. The model may be caused to fail in various failure modes.

3. The entire model can fit in test facilities, such as wind tunnels.

4. Scaled-down loads are within ranges available at test facilities.

The test loading system must reflect the nature of the actual loads superimposed on the structure. For example, ultimate static load tests cannot represent failure due to fatigue. A uniform load may not satisfactorily represent a series of concentrated loads. The most commonly used test loading methods used in current practice are:

1. Discrete loads applied by hydraulic rams (advantageous for quick load release possibility).

2. Dead load, either concentrated or uniformly distributed to simulate materials, such as steel billets, water, sand, gravel, concrete block. (Except for water, most dead loads cannot be removed quickly during test.)

3. Vacuum loading applies uniformly distributed vacuum load on the structure to be tested. (Advantageous for releasing load; however, large vacuum pumps and

bracing systems for other parts of the structure may be costly.)

4. Eccentric fly wheels in pairs with variable speed to provide cyclical vibrating or seismic dynamic loads.

5. Loaded vehicles of known wheel-weight distribution either at slow speeds or used to apply impact load by rolling over bumps on the structure (often used to observe dynamic response of frequency and damping).

6. Wind tunnel testing to study both static and dynamic wind load effects, flow direction, and pressure distribution, and cases of aerodynamic instability.

5.4 DOCUMENT SEARCH

The background of the design, construction and service life of a structure can be reconstructed through the acquisition, assessment, and analysis of the body of documents that are generated during the history of that project. Section 2.6, ("The Document Search") describes the various documents that should be collected and assessed, including those documents that provide specific information with regard to:

1. the history of the site, the design effort and construction of the facilities

2. the condition of the facility at the time of failure, including environmental effects

3. the usage of the facility at the time of failure relative to its service life

4. identification and responsibilities of the various parties

5.5 LITERATURE SEARCH

The search, acquisition, and review of published works dealing with similar failures and failure investigation will aid the investigative team in establishing background data on similar failures and will assist in establishing potential failure hypotheses. As discussed in Chapter 2.8 ("The Literature Search"), these guidelines provide a bibliography which will assist the forensic investigator in initiating a thorough literature search. The bibliography is divided into divisions including a subdivision devoted to structural failures and their investigation.

5.6 REVIEW OF THE STRUCTURAL DESIGN

A structural design analysis should be carried out independently of the original design calculations using the as-built geometry, the service loads, and forces due to environmental effects (winds, seismic, etc.).

In developing a hypothesis about the cause of failure, the investigator must evaluate the original design of the structure to assess and identify incidents of error or omission in the development of the design and the construction documents. A systematic checklist of design factors has been compiled to assist the forensic investigator in developing a hypothesis. Hypotheses that are identified in the profiles of failures by projects and structural type will aid in identifying the particular design functions that contribute to the failure.

The structural designer considers the following factors at some stage of his design:

1. vertical loads (static/dynamic), including dead, live, and roof loads

2. horizontal loads (static/dynamic), including wind, seismic, and those due to stored materials, earth and water pressure

3. construction loads (static/dynamic)

4. cyclical loads that generate structural fatigue

5. impact loads from unusual but possible events

6. exposure of structure to changes in temperature (other than fire), and moisture and their effects on materials

7. exposure of structural members to fire or explosion

8. response of structure to volume changes associated with environmental factors, including forces and displacements

9. soil/structure interaction, including settlement, subsidence, slippage, and differential movements

10. overall structure and element stability with appropriate factors of safety

11. vibration characteristics and sensitivity of structure to hydro/aerodynamic effects

12. structural life as related to maintenance and operational requirements

13. material selection and member sizing within allowable stress, strength, deflection, and serviceability limits.

14. provisions for quality control of materials, methods, and equipment.

Independent structural analysis should be carried out to evaluate the behavior of the structure before and during collapse. These may include static, dynamic, and stability analyses assuming linear or nonlinear material behavior. Depending on the type of structural system and its complexities, inelastic analyses may go beyond the current state of the art or be prohibitive from a cost standpoint. In any case, the goal of a comprehensive independent analysis is to determine what behavior should be anticipated given the structure:

1. as designed

2. as built, including the variations in placement from design and material properties as measured during construction

3. as existed just before failure, material properties as measured from samples from collapse, and other conditions before collapse

5.7   DESIGN DOCUMENT REVIEW

Review of design drawings, specifications, and shop drawings is generally useful in determining a potential breakdown in the communication process between design intentions and contractor's intended fabrication methods.

Review of design calculations does provide valuable information. A knowledge of the designer's assumptions, loading combinations, and anticipated stress levels can provide needed insights for understanding the complexity of the structure and its vulner bilities. The investigator must be very careful in interpreting the meaning of design errors. His or her evaluation of the design must examine the actual behavior of the structure, not just its conformance to code requirements and standard practice applicable at the time of the design. Design errors should be evaluated in terms of their overall effect on performance or collapse. In many cases, errors in member sizing do not significantly affect the structure's performance. More frequently, assumptions relating to the following are suspect:

1. degree of fixity of connections
2. absence of assumed external reactions, or support conditions different from assumed conditions
3. material volume change effects
4. unanticipated loads or load combinations

The results of this detailed review of the design and structural analysis would indicate whether the hypotheses being considered are supported, rejected, or remain unaffected by these analytical studies.

5.8   FAILURE PROFILES

To assist the forensic engineer in identifying potential causes of failure, the forensic investigator is provided with an overview of classic structural failures by type and project. A knowledge of past failure causes and their characteristics can establish a listing of

-121-

potential failure hypotheses for consideration that may be expanded as the investigation advances.

This section consists of a listing of failure causes that will aid the investigator's knowledge of failures, and provide a starting point for further research and development of solutions.

5.8.1 <u>CHARACTERISTICS OF FAILURE BY PROJECT TYPES</u>

This section will discuss common failures classified by function relative to the type of project.

The failures are identified as they are related to specific types of projects, such as bridges, tunnels, multi-story structures, etc.

Structural systems, materials, connections or foundations that are actually utilized in each project type, and their history of failures are discussed under failure profiles entitled, "Structural Systems". Following is a history of failures of project types that may assist the forensic engineer in developing hypotheses of failure.

5.8.1.1 <u>Bridge Failures</u>

The failures of bridge structures may be related to the following specific design factors:

5.8.1.1.1 <u>Design Factors Relating to Bridge Failures</u>

1. span length

2. heavy loading pattern with impact and cyclic loads

3. tendency toward in-creased loadings over lifetime

4. exposure to environ-ment, including tem-perature, wind, mois-

ture, and aggressive
chemicals with the en-
tire structure being
open to these effects
without cladding

5. variety of soil con-
ditions for founda-
tions, including hy-
draulic effects

6. exposure of structural
elements to damage from
vehicles

5.8.1.1.2 <u>Past Failures of Bridges
Unrelated to Structural
Types</u>

Failure of bridges that
are not related to
structural systems, but
are related to factors
contributed by environ-
ment or material fail-
ures, include the
following:

1. scour of foundations
leading to overall col-
lapse

2. fatigue, brittle frac-
ture, and stress cor-
rosion cracking in steel
members

3. collision of vehicles
under bridge with pier
and girder, train de-
railment,shifting train
loads, collision of ve-
hicles over bridge with
truss member or guides,
collision of ships with
pier or superstructure

4. corrosion of deck re-
inforcing due to sal-
ting of surface result-
ing in distress of sur-
face

5. corrosion of floor system, allowing vehicle to
   crash down through the
   deck

6. flooding reaching bot-
   tom of bridge and causing unanticipated horizontal loadings (very
   frequent cause of failure of older timber
   bridges)

7. failure under seismic
   load

8. failure of components
   due to abutments and
   piers

5.8.1.2   Dam Failures

Dam failures are unique because of
their specific structural systems.

5.8.1.1.1 Past Failures of Dams

1. horizontal loading a-
   ssociated with the head
   of water stored behind
   the dam, the effect of
   water pressure on dam
   materials, over-topping

2. erosion of foundation
   from piping

3. subsidence and foun-
   dation movement

4. damage by frost and ice
   pressure

5. insufficient foundation

6. war breaching (acts of
   war)

7. sliding and uplift combined

8. unequal temperatures a-
cross dam cross sec-
tions

9. dynamite blasting near-
by

10. wave action and poor
energy absorption char-
acteristics

11. seismic loadings

12. silting of reservoir
causing greater loads
than assumed during de-
sign

12. slide of upstream lake
shore into reservoir

13. loss of shear capacity
between concrete and
rock or soil surfaces

5.8.1.2.2 <u>Embankment Dams</u>

Embankment dams generally
are composed of earth
materials and/or rock from
sources near the site, and
are sometimes used with
impermeable facings of
concrete, brick, and clay
and cores of concrete or
clay. Sometimes earth and
rock dams have timber
cribs. The features that
make the earth dams
different from other dams
are materials whose
strength properties vary
depending on moisture
content and internal pore
pressure. The construction
and use of the structure
subjects it to variations,
such as permeability,
density, and shear
strength. The primary phenomena or cause associated

with embankment failures are:

1. inadequate spillways and resulting topping of the earth fill

2. rapid draw down and slide failure due to slope instability

3. leaks developing in size into washouts, seepage, slides

4. piping erosion

5. sloughing of upstream clay face

6. saturation and defecttive drainage

7. failure along outlet pipes

8. settlement of puddle core

9. unsuitable material and upstream slope slip

10. breach at intake device

11. unsatisfactory compaction

5.8.1.2.3 <u>Gravity Dams</u>

Gravity dams can be constructed of masonry and concrete. The primary causes of failures or distress have been associated with:

1. failure at construction joints

2.   erosion in foundation

3.   footings too shallow, leading to scour and blowout of foundations

4.   poor material quality

5.   material degradation

6.   horizontal shear of rock foundations

5.8.1.2.4 **Arch Dams**

Arch Dams have plane stresses, in that full design load is very likely to exist and reaction to thrust is 100 percent dependent on the soil and rock conditions. Observed failures or dis-tress were associated with:

1.   structure's ability to adjust to existence of slab plane (in rock under thrust abutment) that was unnoticed in soil investigation

2.   cracking from vibration of penstocks

3.   cracking due to in-adequate grouting of contraction joints

5.8.1.2.5 **Buttress Dams**

Buttress type dams are unique dams due to the fact that horizontal forces are transferred to rock foun-dations using the vertical component of the water pressure in analyzing the dam for sliding and over-turning. Damage to these structures has been caused by:

1. frost damage of abut-
   ments due to alternate
   free/thaw cycles (24)

2. poor masonry bases for
   multiple concrete arches and failure of
   concrete joints

3. cracking of abutments

4. porous concrete

5. undermining of one
   steel bent leading to
   progressive collapse of
   multiple steel arch

5.8.1.3   **Tunnel Failures**

As a structural class, tunnels
are unique in their diversity
because they may be bored
through dense homogenous rock,
in which case the rock may act
as the structure, or they may be
advanced through soft ground
with the use of shields,
linings, and compressed air
techniques. Distress in completed tunnels is often
indicated by:

1. track misalignment in
   railway tunnels

2. moisture infiltration

3. cracking, both longitudinal
   and transverse in lined
   tunnels

4. elevation changes

5. cross section shape changes

Failures in the tunnels may be
associated with the following
causes:

1. distress in linings due to chemical attack (sometimes associated with smoke and moisture)

2. increase in overburden pressure due to increased moisture content of overburden

3. deterioration of linings due to free/thaw cycles

4. vehicular impact damage

5. cracks and corrosion of steel in pressure tunnels due to high hoop stresses

6. failure of filled-in construction shaft due to rotting of timbers of that shaft and overloading of wet sand on brick tunnel arch

7. failure of a concrete-lined spillway due to cavitation at a pump in the lining surface

8. seismic loadings causing geologic slip and shearing distress in tunnels

9. coal mining leading to subsidence under a tunnel

10. inadequate seal against water in water-saturated soils

5.8.1.4 Multi-Story Buildings

High-rise buildings can be classified as structures with a height to least depth ratios of more than two.

Several causes of failures are unique to high-rise buildings. Following is a list of localized and overall

-129-

failure unique to multi-story buildings:

1. localized wind pressures of extreme magnitude causing curtain or panel wall failures

2. airborne objects from adjacent roofs causing window damage

3. overall foundation failure allowing tipping of building

4. large accelerations due to oscillary behavior of overall cantilever action in steady state wind loading

5. progressive collapse due to inadequate shoring, or concrete strength problems in highrise reinforced concrete flat plate construction

6. progressive collapse due to lack of strength, inadequate venting, lack of alternate load paths in precast concrete modular housing

7. failures of the main frame due to seismic loading

8. localized failure due to shortening of columns and facade expansion

9. failure of columns due to loss of foundation support

5.8.1.5 Industrial Buildings

Industrial buildings usually have tall columns, are very large structures in plan dimensions, and may be relatively light construction for economic reasons. Unique failures of these types of structures are as follows:

1. damage to cladding due to high winds or tornadoes

2. ponding of water in flexible, lightweight roof systems

3. failure of structural supports of moving cranes

4. breakdown of slab on grade due to combined effect of concrete shrinkage, high abrasion and impact loadings, inadequate slab thickness, and subgrade compaction, edge, and joint details

5. accidental explosion or purposeful bombing of industrial building

6. uplift of roof structure due to high winds

7. in low temperature, storage buildings roof failure due to build-up of condensate in insulation

8. vibration due to equipment

9. failure due to excessive loading (warehouses)

5.8.1.6 Storage Tanks

Tanks are unique due to the nature of the loads to which they are exposed; failures when reported are related to these loads. Common causes of the failures are:

1. variations in geometry from design assumptions

2. pressure build-ups due to the tank acting as a closed vessel

3. inadequacy of tanks to resist ring tension due to hydraulic head

4. variations in wall continuity around tank

-131-

5. settlement of ring beam foundation

6. material deterioration

Other design and construction aspects of prestressed concrete tanks that are cited as being causes of failures or problems are:

1. bursting in ring steel due to loss or prestress

2. relaxation of steel relative to concrete

3. difference in moisture content, and variance of temperature of inside as compared to outside of tank wall

4. poor design and construction of manhole and pipe openings

5. restraint at base due to rigidity of base to wall

6. voids in mortar due to poor wire spacing

7. bunching of wires sprung around openings resulting in poor mortar coverage or wire

8. poor bond of pneumatically placed cover to core wall or previously placed material, especially if winter conditions prevailed during placement

9. inadequate thickness of cover

10. improper use of pneumatically placed mortar

11. notching of wire during stressing

12. splices of dissimilar material

13. corrosive environment in concrete

14. corrosive environment caused by leakage

15. excessive delay in applying cover coat

16. welding near wire or use for a ground

17. permanent over-stressing of wire

18. inserts in contact with dissimilar metal

19. contact of wire with current-carrying elements

20. winter conditions

21. horizontal cracks and control joints that move and continually work joints through stress reversal cycles

22. localized restraint of walls to volumetric changes by improperly attached appurtenances

23. improper curing of cover coat

5.8.1.7 Chimneys and Stacks

The geometry of chimneys and stacks, are unique in both their tallness and their small cross-sectional areas. This makes them susceptible to:

1. severe thermal stress differentials through their thickness

2. extreme environmental conditions

3. severe chemical exposure on linings

4. high wind exposure and seismic effects as free-standing canti-levers if not guyed or braced

5. sensitivity to differential foundation settlements

6. excessive deflection or swaying of reinforced concrete chimneys over 300 feet in height due to resonance associated with wind eddies.

Self-supported steel stacks have been reported to be particularly vulner-able to excessive vibrations due to periodic vortex discharge and the forces associated with them. Some of the reported vibrations of stacks resulted in localize plate buckling and failure.

5.8.1.8 Guyed Towers

Guyed towers have several unique features. These are:

1. exposure to environment, in-cluding wind, temperature, mois-ture, ice, etc.

2. typically low or no factor of safety against buckling if one of the guys is assumed missing

3. extreme slenderness and potential vibration of components

Although several towers have failed during construction and causes are probably related to the three unique features of design stated above, most failures have occurred during very high wind conditions or ice storms.

5.8.2 FAILURE PROFILES CLASSIFIED BY STRUCTURAL SYSTEMS

The profiles of failures that are related to specific structural systems are generally identified by the geometrical configuration of the system, rather than by the materials of construction that are utilized in the system. There are exceptions, e.g., concrete is always used in flat slabs and steel cables are utilized in suspension structures.

Following are failures that are endemic to specific structural systems.

5.8.2.1 Arched Frames and Rigid Frames Structural Systems

Arched frames and rigid frames are similar and unique as structural systems in that under vertical load their reactions are inclined rather than vertical. Mixes of the two are predominant in actual construction. Cross-sections are designed for thrust, moment, and shear. Tension exists in members only if the line of pressure is outside the kern of the section.

Location of hinges, ties, and points of fixity are critical to design assumptions, as are cross sectional dimensions, material properties, and loadings. Failure and distress in arches and frames have resulted from:

1. change in geometry due to excessive volume changes

2. loss of horizontal thrust at abutments or supports

3. settlement of supports

4. lack of bracing for lateral stability for steel frames

5.8.2.2 Trussed Structures

Trusses are generally designed as structures with ideal pin joints, but normal construction/fabrication of a frictionless pin is virtually impossible. Variations from this idealized frictionless pin may have been responsible for failure unique to trusses. Failure can also result from:

1. member axes with eccentric intersections that induce excessive moments in members designed for simple axial loads

2. fabrication of members at gusset plates that develop fixity

3. addition of secondary members changing the distribution of moment in the member from that assumed in design

4. lateral restraint at locations assumed in design bearings where forces were assumed to be small

5. lack of adequate support for compression

6. lack of lateral bracing to transmit wind forces on truss and live load to end supports

5.8.2.3 **Suspension Structures**

Suspension structures assume versa-tile geometric configuration and have been used in bridges, roofs over large enclosed spaces, and tent structures.

Two primary types of suspension systems are used in bridges: catenary shapes and cable stayed. Catenary shaped suspension bridges are unique because of their lack of redundancies and their susceptibility to aero-dynamic excitation. Failure of sus-pension bridges have been caused by:

1. aerodynamic excitation leading to complete collapse

2. loss of a critical element such as an eyebar in the main tension eyebar chain, which leads to overall collapse due to lack of a

secondary load path of sufficient
strength

5.8.2.4 Suspension Roof Systems

Suspension roof systems typically
employ the use of relatively heavy
concrete decks, double opposing
layers of radiating cables, or
saddle-shaped rods of opposite
cur-vature cables to dampen the
tendency to flutter.

5.8.2.5 Long Span Structural Systems

Generally, the types of roof
systems included in the category
might include all of the following:

1. one-way slabs, beams, and girders
   with simple columns (these
   systems may act as long spans
   depending on many factors, such
   as deflection behavior)

2. trusses (ratio of span to depth
   being the measure of longness)

3. arches, vaults, and suspension
   systems

4. doubly curved shells, membrances,
   space frames, air-supported
   structures

5. folded plates

6. flutter of suspension structures
   and membranes

Failure and distress unique to long
span structures have been associated
with:

1. lack of redundancy and failure of
   a key element

2. ponding of water over folded
   plates and saucer-shaped roofs

3. progressive failure in highly
   repetitive systems without iso-

lating joints (sometimes incorporated in expansion joints)

4. large unstable sections requiring
temporary bracing during construction with no redundancy or
tie to the main stable structure

5. large deflections, rotations, and
lateral movements incompatible
with cladding, electrical, mechanical, and ventilating systems

6. over-stressing due to forcing of
ill-fitting components during
construction

7. lack of camber resulting in excessive ponding due to deflection

5.8.2.6 Continuous Framing Systems

Combinations of slabs, beams, girders, and columns (all with thorough
joint continuity) are a type of
system that seldom collapses, even
when under-designed. Failures as
significant as the loss of beams and
columns have been known to result in
localized failure only. The most
common distress of this type of
system relates to serviceability:

1. deflections exceeding capacity of
finish materials to accept movements

2. vibrations under heavy equipment
loadings (only with high span to
depth ratios)

3. cracking concrete floors leading
to excessive maintenance problems
where concrete is used as a wearing surface under high abrasion

5.8.2.7 Flat Slabs

This is a two-way system of reinforced concrete of uniform thickness
throughout, and is unique in that

optimum design results from mini-
mizing the thickness of the slab.
This attempt to reduce slab thickness
results in high punching shear
stresses in slabs at columns. Failures have resulted from:

1. punching shear at interior columns due to overloading slab when
   design concrete strength has not
   been achieved (most common during
   construction)

2. combined bending, torsion, and
   punching shear

3. excessive deflections resulting
   from construction methods and
   loadings, material properties, or
   slab thinness

4. inadequate attention to volume
   changes resulting from creep,
   shrinkage, temperature change,
   and restraint of those effects

5. inadequate reshaping during construction resulting in distress
   of overloaded slabs at a lower
   level.

5.8.2.8   **Multi-Story Rigid Frames**

Rigid frames are unique structural
systems in that applied loads are
distributed among several members in
the form of bending, shear, axial
forces, and torsion. As the number
of redundancies increase in a
structure, such as a rigid frame, its
ability to redistribute forces from
over-stress and individual members is
increased. Even if individuals
members or restraints are removed,
the structure will often remain
stable and stand-ing, although
possibly over-stressed. For this
reason, few overall fail-ures of
rigid frames have been reported. However, brittle failures have been
associated with large welds in highly
restrained, heavily loaded connec-

tions even where design, materials, and construction techniques have appeared adequate by present standards.

5.8.2.9 Thin Shells and Membranes

Thin shells or membranes are defined as curved or folded continuous material whose thickness is small compared to its other dimensions. Two ways in which they may be classified are type of curvature, or method of generation. Singly curved surfaces are usually cylindrical or conical. Shells of positive double curvature are domelike, and shells or negative double curvature are saddle-like. Curvature of these structures results in predominantly in-plane forces resisting loadings. Distress or failure of these roofs has been associated with:

1. heavy concentrated loadings

2. instability associated with very flat or very thin shells or shells of low modulus concrete

3. creep of concrete

4. failure at boundaries of the primary shell or membrane system

5. changes in geometry due to any cause

5.8.2.10 Cantilevers

Cantilevers are unique in that they have no redundancy. Failure and distress of cantilevers as a structural type of element have been associated with:

1. inadequate lateral support of compression flange of rolled shapes or compression face of thin members

2. intolerable oscillatory motion due to wind

3. overloading

4. inadequate strength or behavior of members framing into same joint, especially back-spans in continuous beam systems

5. connection failures

6. deterioration due to environmental exposure

7. excessive deflection due to modular of elasticity of concrete being low

5.8.3 FAILURE CASES CLASSIFIED BY MATERIALS

Following are common causes of failure that are caused by faulty materials of construction:

5.8.3.1 Steel

The mechanical properties of steel depend primarily on the chemical composition rolling processes, and heat treatment techniques employed. Other factors that may give different results for the same grade of steel are rate of loadings of specimen, condition, and geometry of specimen, cold work, and temperature at the time of testing. For the purposes of steel, they may be categorized as carbon steels, high strength, and high strength low alloy steels, and quenches and tempered alloy steels.

Specific American Society for Testing and Materials (ASTM) designations are provided for the three categories described above and for light gauge cold formed members, castings forgings, rivets, bolts, and weld filler metal. Wire, strand, cable, and rope, including their coatings, are described in the standards established by the Wire Rope Technical Board.

The most common causes of failure of open steel rolled shapes or fabricated sections are:

1. lack of lateral support of compression flange of beams

2. lack of lateral support for columns

3. failure of welds

4. web crippling or buckling

5. lack of consideration to stability of members and overall structural system

6. lack of consideration to torsion effects

A less common cause of failure, sometimes associated with sudden overall collapse, is brittle fracture. Some factors associated with brittle fracture that often occur in combination are:

1. high tensile stresses
2. high carbon content
3. rapid loading rate
4. presence of notches or flaws
5. repeated cycles of loading
6. corrosion
7. highly restrained welded connections

5.8.3.2   Concrete

The mechanical properties of concrete principally depend on the chemical composition of cement, the size, grading and quality of the large and small aggregate, quality of water, mixing, placing, finishing, and curing conditions, and environmental con-ditions such as humidity, exposure to aggressive chemicals, and tem-perature. The process of

-142-

hydration of the cement that bonds
the various mineral fragments
together into a compact whole
controls many of the mechanical
properties of concrete. The more
permeable the concrete is, the more
easily it will allow entry or escape
of moisture, and/or chemicals that
affect its components, and hence
hinder its durability.

Causes of failure or distress of
concrete, as a material, have resulted from either one or a number of
the following in combination:

1. repeated freeze/thaw cycles

2. excessive shrinkage or expansion
    of its components or of it as a
    whole

3. chemical attack resulting in loss
    of volume (for example from
    sulfates or chlorides)

4. strength degradation due to high
    alumina content combined with
    high temperature and/or high
    water cement ratio

5. ingress of water and air usually
    in presence of chlorides, to
    permit corrosion of reinforcing

6. bacterial fermentation in void
    spaces leading to bursting pres-
    sures and disintegration

5.8.3.3   **Masonry (Brick, Stone, Terra
           Cotta, and Concrete Masonry**

           Concrete masonry tends to expand
           and contract with increases and
           decreases in moisture content.
           Burned clay products have irre-
           versible moisture expansion. For
           concrete masonry units, shrinkage
           after construction is a major
           factor. Due to the composite
           nature of masonry construction,
           the performance of the composi-

-143-

tion of materials used in masonry construction is affected by the mortar connecting the masonry units.

5.8.3.3.1 <u>Failures and Distress of Masonry have Resulted from</u>:

1. freezing of confined water in joints and in units resulting in cracking and spalling

2. expansion of masonry foundation due to moisture content and subsequent shrinkage resulting in distress to the portion of the masonry extending above grade

3. thermal movement of masonry described above, which tends to put tension in the masonry and leads to cracks

4. restraint in the expansion of masonry parapet walls causing bowing of the walls. This expansion may aggravate existing conditions in parapet due to the two faces of the parapet being exposed to extremes of moisture and temperature

5. movement of masonry encased structural elements causing the masonry to crack

6. movement of concrete floor slabs, which

are built into masonry walls at corners, shrinkage and bending resulting in rupture of masonry walls.

7. restrained differential thermal movement in cavity wall construction that results in vertical cracks at corners

8. restrained expansion of walls running in the same direction, resulting in cracking at offsets in plane of wall

9. unaccommodated shortening of high-rise concrete building frames under vertical loading, resulting in bowing and cracking of masonry facing walls

10. foundation settlement causing cracking, which varies in width from top to bottom

11. corrosion of metal ties and structural steel members

12. leaking of walls improperly designed, built, or maintained

13. buckling and delamination of masonry pavements caused by the proper expansion joints

14. staining of masonry due to improper se-

lection of mater-
ials, or design or
construction error
or omission

15. indiscriminate ap-
plications of im-
proper coating

16. lack of accommo-
dation of differ-
rential movement be-
tween materials
causes cracks.
Cracks in turn
cause water leaks,
which lead to most
building problems

5.8.3.4 Wood

Wood is a unique construction material in that it is composed of an
organic cellular grained structure,
and it is significantly anisotropic.
Wood is often assumed to be ortho-
tropic in nature, with the three
principal elasticity directions taken
in the longitudinal radial and tan-
gential directions in the tree element.   Due to its variation in
properties with direction, three
Young's moduli, three shear moduli
and three Poisson's ratios would be
required to specify its elastic
properties. Typically, however, the
most important elastic constant used
in design is the modulus of elasticity along the grain.

Many failures are associated with the
variation of strength properties
associated with:

1. knots
2. moisture content
3. grain, or sloping grain
4. density
5. shakes
6. splits
7. checks

Other types of failures have been associated with the deterioration and degradation of properties due to:

1. repeated saturation and drying

2. attack of fungi of a variety of species through enzymatic processes, degrading and decomposing its cellulose and lignin

3. attack by wood-destroying termites and beetles

4. exposure to heat and fire

Finally, the most common failures of sizeable timber structures are associated with:

1. volume changes associated with drying shrinkage

2. connection details

3. inadequate maintenance such as tightening bolted connections

4. failure to prevent degradation due to alternate wetting and drying

5.8.4   **FAILURE CAUSES CLASSIFIED BY CONNECTION TYPES**

Overall collapses resulting from connection failures have occurred only in structures with few or no redundancies. Where low strength connections have repeated, the failure of one lead to failure of neighboring connections, and a progressive collapse occurs. The primary causes of connection fail-ures are:

1. improper design due to lack of consideration of all forces acting on a connection, especially those associated with volume changes

2. improper design resulting in abrupt section changes and stress connections

3. insufficient provisions for rotation and movement

4. improper preparation of mating surfaces and installation of connections

5. degradation of materials in a connection

6. lack of consideration of large residual stresses resulting from manufacture or fabrication

7. corrosion, especially due to contact of dissimilar metals

5.8.4.1 Steel-to-Steel Connections

5.8.4.1.1 Rivets and Bolts

The writers are not aware of any structural failures resulting from failures of rivets or bolts of common structural steel grades, other than those associated with connections not being completed, fatigue cracking through holes, and corrosion of grossly undersized fasteners. However, high strength alloy steels or heat-treated bolts used in special high-stress applications may be susceptible to brittle fracture.

In addition, in recent years, failures of connections have occurred due to the use of "counterfeit" bolts. These "counterfeit" bolts are generally manufactured overseas and are fabricated to appear very

similar to the authentic
high strength bolts.

5.8.4.1.2   **Welded Connections**

If the proper design,
specification, material,
and techniques are used,
welded connections
typically per-form as
anticipated. However, in
plate girders, certain
changes in section at
splices, vertical hori-
zontal web stiffeners, and
connections of transverse
secondary framing are prone
to fatigue cracking.

5.8.4.1.3 **Pinned Joints**

These joints are encoun-
tered in structures such as
trusses where ends of members are assumed to be free
to rotate without developing moments. Transfer of
forces at the connection is
from the end of the member
bearing on the pin, shear
on sections within the pin,
and transfer through
bearing to adjacent mem-
bers.

Failures associated with
pinned connections have
been due to:

1. brittle fracture of the
   member at the pin hole

2. corrosion at the
   connection

3. loss of keeper on pins
   and moving of the pin
   laterally out of the
   joint

5.8.4.2 Steel-To-Concrete Composite Connections

Shear connectors are required to positively engage the concrete for composite action. Distress of these connections has been caused by:

1. inadequate weld of connector to the parent material

2. inadequate compaction of the concrete around the shear connector

5.8.4.3 Cast-In-Place Concrete Connections of Precast Concrete Members

The writers are not aware of any collapse resulting from loss of bond of cast-in-place concrete connections to precast members, although there have been cases of excessive laitance at the smooth top surfaces of precast concrete that resulted in delamin-ation at the bonding interface and excessive deflections.

5.8.4.4 Other Connections of Precast Concrete

Several of the common types of connections used in precast concrete construction are listed below, together with their respective causes of distress or failure:

1. steel-to-steel, connections have often been improperly protected, and show distress due to cor-rosion

2. elastomeric bearing pads, if used with excessive thick-ness in both primary and secondary framing members without other lateral sup-

port, have lead to insta-
bilities and collapse

3. coil anchors and rods have
    failed due to inadequate
    thread embeddment into the
    anchor and insufficient edge
    distance, embeddment depth,
    or spacing to develop the
    anchor strength

4. grouted connections using
    Portland Cement mortar, e-
    poxy with reinforcing, bolts
    or post-tensioning across
    the joint, fail due to im-
    proper proportioning of
    materials, preparation of
    surfaces, and lack of ade-
    quate development of the
    reinforcing bars

5. direct bearing of precast
    concrete members on rigid
    concrete supports have
    failed due to rotation and
    sliding at the corners,
    resulting in stress concen-
    trations

6. dapped ends have failed due
    to inadequate reinforcing or
    improper placement, or
    inadequate provisions for
    forces due to volume changes

5.8.4.5 Post-Tensioned Concrete

Many failures have resulted from inadequate corrosion protection of the
tendons or anchorage by:

1. incomplete greasing or loss of
    grease around tendons

2. incomplete grouting of tendons,
    either due to poor field tech-
    niques, quality control, or
    sedimentation of the grout

3. seepage of water into anchorages
    along cold joints between an-

-151-

chorage block-out patches and parent concrete

Other collapses have been associated with:

1. notches in tendons due to methods of pre-stressing tanks

2. seismic loading on wedge type anchors in unbonded systems

5.8.4.6 **Monolithic Concrete Member Intersections**

These intersections, whether in precast or cast-in-place concrete, exhibit high strength due to their homogeneity and continuous reinforcement. Failures in these intersections have resulted from:

1. discontinuity in reinforcement due to inadequate embeddment or splices

2. high shear and bending stresses at column/beam joints sometimes associated with opposing post-tensioning forces

3. punching shear failures in flat plate construction

4. inadequate cover and protection of the reinforcement from corrosion

5. corrosion of reinforcement and prestressing hardware at cold joints that are inadequately protected

5.8.4.7 **Masonry Concrete**

The failure of joints between masonry and concrete are often the result of:

1. corrosion of ties

2. rigidity of ties and resulting overstress of tie anchorage due to volume change effects

3. inadequate control of construction tolerances

4. improper anchorage of shelf angles

5. design of shelf angles for strength but not deflection

6. improper design and control of sealant joint design and construction

7. improper design and installation of wall flashing

5.8.4.8 Timber Fasteners and Adhesives

Common connections are listed below with their associated distress and failure causes:

1. Nails, bolts, and screw lose withdrawal and lateral resistance if subjected to wetting and drying cycles; heads may rust off due to corrosion.

2. Connections, including bolts, split-rings, toothed-rings and shear plates used in green wood can loosen as a result of shrinkage.

3. Glued joints have failed due to extended high temperatures or high moisture environments.

5.8.4.9 Expansion Type Connections

The most common cause of failure of expansion type connections occurs when the joints freeze, leading to restraint of members that were designed to move freely and rotate. The following are typical expansion

connections and their associated
causes of failure:

1. Sliding plates of equal areas
have large contact areas and can
freeze, often leading to distress
of the corbel or ledge.

2. Elastomeric bearings, if improperly sized, can fail by plastic
flow at high stresses, or by
rolling if shear distortions are
large compared to the depth.

3. Self-lubricating bearings such as
two steel plates with a bronze
lubricated plate between them can
seize if not kept free of dirt.

4. Roller and rocker bearings, and
pedestals and shoes have developed distress in structural
elements when actual movements
are greater than anticipated or
they become clogged and inoperative due to lack of maintenance or poor design.

5. Slab or deck expansion joint
covers have been known to foul
due to corrosion, collection of
debris, or build-up of ice, which
results in restraints of con-
nected members.

6. Expansion anchors suspending
intermediate spans in continuous
girder bridges have failed due to
restraint caused by welded details preventing free rotation
and translation.

Brittle fracture and corrosion of
prestressing steel in concrete
structures have also been the subject
of extensive research.

5.9   SYNTHESIS OF THE INVESTIGATION

As data and findings are collected from the investigative team, and efforts are made relating to document search, literature search, and site studies, testing programs and design analysis, the process of synthesizing the body of information into a failure hypothesis will be implemented.

The principal investigator will synthesize the findings of the investigation team, communicate the results to the team, and develop a program that will establish the most likely cause or causes of failure.

The synthesis of the findings of the investigation will culminate with the development of failure hypotheses. Following is a listing of systematic appraisals of data that will lead to the efficient development of failure hypotheses. Each listing will include reference to additional sections that will guide the investigator in that task.

5.9.1 DEVELOPMENT OF A HISTORY OF THE DESIGN AND CONSTRUCTION OF THE FACILITY

Use the material gathered in the document search (Chapter 2.0) to provide an overall narrative history. Refer to Chapter 3.0, "The Report," to review historical data required for the report.

5.9.2 DESCRIPTION OF SITE AND SERVICE CONDITIONS AT TIME OF FAILURE

Use the materials gathered in the document search (Chapter 2.0), eyewitness accounts (Chapter 5.0), and field observations (Chapter 5.0) to develop a narrative account of the salient conditions prevailing prior to the failure.

5.9.3 DESCRIPTION OF THE FAILURE

Use the eyewitness accounts (Section 2.5), field observations (Section 2.5), test results (Chapter 5.0), and review of field data (Chapter 5.0) to describe the failure.

The account should also include a description of the "Trigger Mechanism" that initiated the collapse and the prevailing condition that acted on the trigger mechanism and caused the structure to be overstressed or unstable.

5.9.4 REVIEW OF THE DESIGN OF THE STRUCTURAL SYSTEM

To aid in the development of a comprehensive review of the design of the structural system relative to its original intended use, and its use at the time of failure, use the guidelines contained in Section 5.6 to assist in developing a comprehensive structural design review.

5.9.5 DEVELOPMENT OF STRUCTURAL FAILURE PROFILE

The development of a narrative and graphic overview of the failure profile can be carried out using Section 5.10 as well as the results of Section 5.8.2 and 5.9. Section 5.11 describes classic structural failures with regard to project types, structural systems, materials of construction, and types of connections.

5.9.6 CATEGORIZATION OF THE FAILURE

Using the analysis derived from Sections 5.8.2., 5.9 and 5.10, the failure can be classified. The information included in Section 2.3, "The Investigation," may be helpful.

5.9.7 DEVELOPMENT OF DESCRIPTION OF THE FAILURE

The data developed during visual observations in the field and the test program can be used to recreate the failure scenario.

The observations relating to the overall movement of the structure or the behavior of the collapse, the description of member distress and the fracture surfaces, and the condition at the time of failure will provide background information leading to the development of the failure scenario.

5.10 DEVELOPMENT OF FAILURE HYPOTHESIS

During and after the synthesis of the information gathered in the course of an investigation, hypotheses on the possible cause or causes of the failure will be advanced. The various sections of this guideline including 2.0 and this chapter are intended to aid in the formulation of failure hypotheses.

The principal investigator should manage the team and its synthesis of information to maintain a wide perspective of alternate hypotheses so every possible cause is considered.

The various hypotheses that are developed using these guidelines must be maintained in a written narrative. This will allow the investigation team to use the record in the event that new evidence is uncovered that indicates that a specific hypothesis should be modified, reconsidered, or eliminated. Using these techniques, the several hypotheses generated from the investigation synthesis will be proven or disproven, resulting in a failure hypothesis that can be supported by the principal investigator.

When the failure hypothesis is advanced, the preparation of the investigation report may be initiated using information contained in Chapter 3.0.

PREPARATION FOR THE LEGAL PROCESS

6.1 **INTRODUCTION**

The legal considerations related to a forensic engineering commission illustrate the reality that the engineering investigation of a failure inci-dent is a fact-finding mission that results in uncovering the probable causes of that failure; as part of the process of identifying the cause of a failure, the party or parties respon-sible for the failure should be revealed. Thus, the product of the investigation, the opinions of the forensic engineer, in the form of written reports, will be subject to judicial review in a legal dispute.

This chapter has been prepared as a guide for the professional who has no experience as an expert witness, and as a forum for the presentation of some new ideas for the veteran. The contents are intended to provide a generalized overview of the legal considerations attendant to a forensic engineering investigation, to explicate the legal process.

This overview of the legal aspects of forensic engineering applies to almost every type of sit-uation in which an engineer is retained to investigate a failure. However, this section is oriented toward a scenario wherein a forensic engineer is retained as an expert witness by an attorney representing either the defendant or plaintiff in a legal dispute.

6.1.1 **Role of the Expert Witness**

The role of the expert witness, when ap-pearing in a legal forum, will be to assist the court as an interpreter of technical matters. This is done by explaining complex technical issues, conveying expert opinions and presenting conclusions about these issues.

An expert in forensic engineering is not automatically dubbed an "expert witness." An individual is not granted the privileges of an expert witness until a legal forum confers that recognition upon a properly qualified professional. When assigned the role of an expert witness, the person is permitted to render opinions on the issue; this opinion will then be considered as evidence.

The acceptance of an opinion as evidence differentiates between the "expert witness" and a "fact witness." The fact witness may not express opinions, but may only report events that were "observed."

The expert should also assist the client in understanding the technical aspects of the case, using observations, analyses, failure hypotheses, and recommendations as a frame of reference.

While remaining objective, the forensic engineer should assist in the evaluation of weak and strong points of the case, participate in the analysis of issues, advise in the preparation of complaints and interrogatories, assist in preparing questions for adversary witnesses, and help to evaluate the equity of settlement offers.

An element of essential importance is the overriding professional principle of honesty. This principle cannot be violated. The opinion expressed by an expert witness during the course of a legal dispute must be the same, no matter who has retained the expert.

In the majority of legal disputes a client is usually unsure of his or her legal position until the forensic investigation is complete, when the cause of the failure is identified, and legal issues have been thoroughly researched. If the case goes to trial and the expert witness appears in court, the expert who has presented his/her opinions in a direct, unbiased manner can give testimony that is unassailable. However, the expert who assumes the role as an advocate of the client's position, slanting his or her opinion to aid the client's cause, has violated the principle of honesty. Thus, advocacy places the expert in a vulnerable position. Should this dishonesty be exposed the complete contents

of the expert's testimony may be regarded as
suspicious. Therefore, when considering a
forensic engineering com-mission, it is
important to determine what role will be
expected by the client. Will the expert be
permitted to carry out this commission
objectively and report the truth? Or will
the client require that the expert witness
assume a subjective role and support the
client's position? The commission should not
be accepted unless the client will permit
the forensic engineer to assume a role that
maintains an objective and unbiased
position. The alternative role may require
the expert witness to commit perjury.

6.1.2 SUMMARY: PREPARATION FOR THE LEGAL PROCESS

In order to fully evaluate the forensic
expert's role in the legal process, the
expert should understand the definitions and
concepts of a dispute resolution. In order
to provide an overview of what may be
expected by the forensic engineer during the
preparation for the legal process a summary
of these elements is contained in this
chapter.

6.2 ALTERNATIVES FOR ENFORCING LEGAL RIGHTS

The professional services provided in the investigative process and the duties expected of the
expert witness are generally the same no matter
what method of formal dispute resolution is
selected by the client. Following is an overview
of the various methods of dispute resolution that
may be encountered.

6.2.1 CIVIL LITIGATION

The Civil Litigation process occurs in a
court of justice having jurisdiction in the
dispute. The procedure for litigation is
mandated by laws that may differ from
jurisdiction to jurisdiction. The court is
presided over by a judge who decides on the
applications of the law and procedure.
These decisions usually are determined by a
jury of lay-persons, or by the judge in the
absence of a jury. The decision of this
court and jury may be subjected to appeals

and retrials. Civil litigation is sometimes
a lengthy and expensive procedure. Many
civil actions are initiated but terminate in
out-of-court settlements. These out-of-
court settlements occur principally when the
magnitude of the losses are understood by
both plaintiffs and defendants. The
settlements account for anticipated money
and time saved by not going to trial.

6.2.2 ARBITRATION

Instead of a judge and/or a jury, as in the
Civil Litigation process, an arbitrator or a
panel of arbitrators may be used in a private proceedings. Arbitrators are selected
by the disputants by process of elimination,
using a list of qualified individuals.

Unlike litigation, arbitration eliminates or
significantly relaxes the discovery process
and is usually limited to the production of
documents to an extent determined by the
arbitrators. The American Arbitration
Association is one of several organizations
that has established rules for the conduct
of the proceedings. They have offices in
major cities across the nation. A local
office may be contacted for further
information.

6.2.3 ADMINISTRATIVE HEARINGS OR SPECIAL INVESTI-GATIONS

Hearings are held by governmental authorities for a number of reasons. Hearings
may be held to resolve disputes or to gather
information or public opinion prior to making an administrative decision, or for inquiries into the integrity of the structure
or the cause of a failure of an engineering
facility.

6.2.4 PRIVATE LITIGATION

This form of dispute resolution requires
both parties to agree to resolve their
differences in a private court that is
created, and paid for, by both parties. A
mutually respected individual, such as a
retired jurist, is chosen as a judge, and
civil litigation rules are followed. This

is frequently referred to as a "mini-trial."
Corporations, insurance companies, law
firms, and individuals often resolve
disputes using private litigation instead of
filing with courts. Several organizations
that serve as private court systems have
been established to serve this need.

6.2.5 MEDIATION

This method of "non binding" dispute resolution is used to resolve certain construc-
tion-oriented disputes. A mediator acceptable to both sides is selected and probes
the disputants to determine their position
relative to "minimum" settlement. He or she
then strives to attain a middle ground that
both can accept comfortably. The American
Arbitration Association will assist in the
mediation process.

Mediation can be used for practically any
dispute.

6.2.6 MANDATORY DISCUSSION, MEDIATION, AND ARBITRATION

This method of dispute resolution has been
used for years to solve labor disputes and
is now being used as a technique for
resolving construction disputes. This method
of dispute resolution requires a mediator
familiar with construction arbitration to
intervene after a dispute occurs. The Deep
Foundation Construction Industry Roundtable
(DFCIR) recently proposed language to be
incorporated in contractual agreements
between the owner and the design professionals, and the owner and the general
contractor and subcontractors and material
suppliers. This document provides for immediate non-binding mediation; and, failing
such a resolution, immediate binding arbitration, all to be concluded within a 100-
calendar-day period.

6.3 THE CIVIL LITIGATION PROCESS

The following discussion is provided as an
overview of the definitions and the methodology of

-162-

civil litigation. Due to the sequential nature of the process, it is possible for an expert witness to start serving at almost any point during the process. Of course, it is beneficial to be commissioned at the initiation. Joining the proceedings after they have begun requires the expert to "catch up." Being contacted late in the process may be an important consideration in deciding whether or not to accept the assignment.

The following is a sequential series of descriptions beginning with the actual incident that promulgated the dispute and testimony through stage I (pleadings), stage II (pretrial proceedings), stage III (trial), and, finally, stage IV (post trial).

6.3.1 THE INCIDENT

Failure or non-performance of an engineered construction eventually leads to a dispute. The disputants may attempt to reach an accord on their own or with the aid of an attorney. If these attempts at settlement are not successful, then the dispute may go into the most popular method of dispute resolution - civil litigation.

6.3.1 STAGE I: THE PLEADINGS

Civil Litigation is initiated when one party files a formal complaint or claims against another party. Following is a series of definitions, offered in approximate chronological sequence, of the salient steps involved in this phase of civil litigation.

6.3.2.1 The Complaint or Statement of Claim

The complaint is also known as "statement of claim," "petition," or "declaration." The complaint (by the Plaintiff) is a statement of allegations made against the other party (the Defendant) and the nature of damage sought; this is normally in the form of monetary awards.

6.3.2.2 The Plaintiff

The party initiating the action with a complaint.

6.3.2.3 The Defendant

The person being served with a complaint.

6.3.2.4 Service

Once the complaint is filed, the defendant receives a summons, sometimes delivered in person by a marshal, sheriff, constable, or other process-server. The summons identifies the defendant and the action being taken against him or her, including monetary damages, the name of the court, the name of the plaintiff, and the name and address of the plaintiff's legal counsel.

6.3.2.5 Default Judgment

The defendant is typically given about 20 to 30 days to respond to the complaint in order to avoid having a default judgment entered. In other words, failure to respond to the complaint within the allotted period would cause the judge to rule in favor of the plaintiff.

6.3.2.6 Appearance

After service of the complaint, the defendant will retain an attorney and make an appearance before the court. This means that the defendant's attorney will submit a formal notice that the summons has been received, and that the defendant is represented by counsel. At this point, attorneys for the plaintiff and the defendant may attempt to reach a settlement.

6.3.2.7 Stipulation

A document filed with the clerk of the court indicating that the plaintiff and defendant have reached a settlement prior to trial is called a "stipulation." The document represents the efforts to settle the dispute between the plaintiff and the defendant. If they do not reach a settlement, the litigation process continues. If the stipulation is filed with the clerk of the court, the case is then recorded as closed.

6.3.2.8 Motion to Dismiss

In conjunction with efforts to reach settlement, the defendant's attorney may argue that the court should dismiss the complaint because the plaintiff does not have a legal right to a judgment in the plaintiff's favor.

6.3.2.9 Other Motions

Various motions may be made to dismiss the complaint; for example, the defendant can challenge the court's jurisdiction over the claim, or of the formal (as opposed to legal) sufficiency of the claim. Motions may be made to strike from the complaint those parts deemed redundant, immaterial, superfluous, or scandalous. These various motions create an early testing of the complaint and the attitudes of the parties toward it.

6.3.2.10 Defendant's Answer

After motions have been made and decisions reached (if the complaint has not been dismissed), the defendant may reveal his or her position with an answer to the complaint. This answer may take the form of a denial, an affirmative defense, a counter-claim, or a combination of these.

1. Denial: A denial typically admits certain parts of the plaintiff's claims may be true, but denies the other parts. This limits the areas of dispute; and if the plaintiff is unable to prove the elements that are denied, the entire case may fail.

2. Affirmative Defense: An affirmative defense tends to indicate that the plaintiff's allegations are true, but explanatory facts have been omitted. Examples are: a claim where the allegations are true but were caused by contributory negligence by the plaintiff, or the statute of limitations has expired.

3. Counter-Claim: A counterclaim alleges facts that could have been asserted by the defendant had he/she wished to file suit. It may be based on entirely different claims and include a demand for damages that could be far in excess of the plaintiff's complaint. In essence, a counter-claim becomes a cross-suit, and the plaintiff must respond, going through the same process that the defendant went through in answering the plaintiff's complaint.

4. Reply: The plaintiff's answer to the counter-claim.

5. Pleadings: If the answer is in the form of a counter-claim, the pleadings consist of the complaint, the answer or counter-claim, and the reply from the defendant. Once closed and submitted, the pleadings identify the only issues that can

be raised at trial, the exception being amendments to the pleadings, which can be made within certain pre-established parameters.

6. Judgment in the Pleadings: After the pleadings have been closed, either party can move for a judgement on the pleadings, whereby the court examines the strength and validity of the various claims.

6.3.3 STAGE II: CIVIL LITIGATION PRETRIAL PROCEEDINGS

After the pleadings stage is complete, civil litigation pretrial proceedings begin. Following is a description of the civil litigation process.

6.3.3.1 Notice of Trial

Either side files a notice of trial requesting the clerk of the court to schedule the suit on the jury or the non-jury calendar. Either side can demand and receive a jury trial, even if the other side prefers otherwise.

6.3.3.2 Pretrial Hearings

Attorneys for each side are required to appear before the judge in chambers to remedy defective pleadings, eliminate superfluous issues and simplify others, agree to which documents are genuine, limit the number of expert witnesses, determine the scope of discovery, and decide if a master should be appointed to obtain information pertinent to the suit.

6.3.3.3 Admissions

Prior to trial, a judge may order an attorney or the attorney's clients to make pretrial admissions, whereby one side admits to the existence of facts that help the other side, or admits that a point made by the opposing side is true.

Admissions can be requested by any party.

6.3.3.4 **Settlement Negotiations**

In order to save time and possibly avoid a trial, a judge may order both sides to engage in settlement negotiations and report the progress or success of these negotiations within a specific time period.

Settlement negotiations are frequently mandatory in many courts prior to trial.

6.3.3.5 **Discovery**

The pretrial process known as "discovery" serves to expedite trial preparation, as well as the trial itself; and requires mutual disclosure of evidence. It eliminates the needless time and expense that would be required by each side to individually obtain valuable facts relevant to the trial if there is no agreement between the parties to exchange information.

During the discovery process, the expert witness should prepare a complete list of documents required for the full investigation and trial preparation. Section 2.6 provides an overview of valuable documents that should be collected as part of the discovery process.

Discovery preserves documentation that may change or disappear for use as trial evidence, and removes the "surprise element" from litigation.

6.3.3.6 Subpoena Duces Tecum

A subpoena duces tecum is issued by the clerk of the court to compel the other party to provide certain documents or other items.

6.3.3.7 Interrogatories

Interrogatories are an additional form of discovery whereby the opposing attorney prepares a written list of factual questions for the other side to answer. The questions must be answered under oath and may later be used during the trial.

6.3.3.8 Deposition

Depositions are a form of discovery in which the opposing attorney asks questions of witnesses. During the deposition the witness is under oath with the proceedings recorded by a court reporter.

The role of the expert witness in the deposition process is important; the expert witness will most certainly be deposed by the other side, and may also be asked to attend the deposition of the opponent's witnesses in order to provide guidance and develop questions for his or her client's attorney.

The transcript of a deposition can be used as evidence in court if the defendant is not available at the time of trial or if answers given in court are different from those given in the deposition.

6.3.3.9 Summary Judgment

The final pretrial procedure is usually a motion for summary judgment to obtain a dismissal of the claim, the counterclaim or both. Unlike a motion to dismiss, it

alleges a case has no merit because
the party bringing it cannot prove
the alleged facts are true or has
not properly based the case in law.

6.3.4 STAGE III: CIVIL LITIGATION TRIAL PROCEEDINGS

When the time for the trial approaches,
attorneys for both sides will answer the
calendar call, and the suit will be assigned
to a courtroom. The judge usually has been
assigned when the complaint is filed. The
following are salient factors in the trial
stage of civil litigation.

6.3.4.1 Selection of Jury

The judge and sometimes the at-
tornies conduct voir dire, a procedure where prospective jurors are
questioned to determine their
qualifications. If a juror's answers indicate a prejudice, a relationship with the parties, a
conflict of interest, or any other
situation meriting disqualifi-
cation, an attorney may "challenge
for cause" and state his reason.
Each attorney is given a number of
peremptory challenges, giving him
or her the freedom to excuse a
juror without giving a reason.

6.3.4.2 Opening Arguments

Once the jury is empaneled, the
trial begins and the plaintiff's
attorney starts the proceeding by
outlining the facts he or she in-
tends to prove. After this summary
the defendant's attorney addresses
the court and provides the court
with a description of the
defendant's case.

The plaintiff's attorney then calls
witnesses for the plaintiff; the
plaintiff always proceeds first at
each phase of the process. Each
witness for both sides is bound by

-170-

the various rules of evidence,
including the following:

1) **Parole Evidence Rule:** This rule
makes inadmissible any evidence
of understanding different from
those entered into when the
formal agreement was est-
ablished.

2) **Relevancy Rule:** This rule holds
that only evidence relating to
issues under dispute as outlined in the proceeding may be
used.

3) **Direct Evidence:** This is
evidence that demonstrates the
existence or truth of facts.

4) **Circumstantial Evidence:** This
is evidence that comprises
indirect proof or disproof of a
fact in question.

5) **Hearsay Rule:** This is evidence
that may be excluded because a
witness quotes what <u>someone
else</u> said about an incident.

6) **Best Evidence Rule:** This rule
holds that the best possible
form of evidence must be
produced for the trial. For
example: If a certain document
is produced as evidence during
trial, the original, if a-
vailable, rather than a photo-
copy shall be used. However,
photocopies may be used in
evidence.

6.3.4.3 <u>**Testimony of the Lay or Fact
Witness**</u>

A fundamental rule of evidence is:
that all lay witnesses may testify
only to matters of fact, and they
may not express their conclusions
on liability or guilt. Conclusions
are to be made only by the judge or
jury.

6.3.4.4 Testimony of the Expert Witness

In contrast to the testimony permitted a lay witness, the expert witness is permitted to express opinions and conclusions, and these statements can be evidence provided the following conditions prevail: The matter on which the expert is testifying is personally known to the expert or has been made known to the expert at or before the trial. In addition, the matter on which the expert relies to form his or her opinions is the same type that any expert in the same field would rely on to form opinions on a similar subject. Furthermore, the expert cannot rely on any matter that he or she is forbidden to rely on by law.

6.3.4.5 Recognition As An Expert

In order for the court to recognize and accept a professional as an expert witness, the attorney who retained the expert will request that the expert recite pertinent credentials. Opposing council may challenge the expert's credentials. This tactic is seldom successful. In most cases, the opposing attorney and the judge will stipulate that the professional is qualified to serve as a witness. It is important that a Curriculum Vitae of the expert be presented to make clear his or her expertise (Chapter 1.0).

6.3.4.6 Direct Examination

Once the expert is accepted by the court, the attorney for whom he or she is appearing begins questioning through direct examination. Under direct examination, the attorney may not ask leading questions, that is, questions that suggest a preferred answer.

6.3.4.7 Cross-Examination

At the completion of direct examination, the opposing attorney
commences cross-examination, asking
questions about testimony made
during direct examination. In
practice, cross-examination is applied in an attempt to cast doubts
upon facts asserted to during
direct examination. This is done
by attempting to catch the witness
in a contradiction or by casting
doubt on the witness's character or
capabilities.

6.3.4.8 Impeaching the Witness

The opposing attorney will test the
recollection, knowledge, and
credibility of the expert. When
the attorney attempts to catch the
witness in a contradiction, it is
called impeaching the witness.
Impeachment sometimes occurs when
the testimony given during the deposition contradicts what is said
at trial.

6.3.4.9 Redirect Examination

The expert's attorney will engage
in redirect examination in order to
correct misinterpretation of answers given during cross-examination, or if the witness's testimony seems to have been im-
peached.

6.3.4.10 Re-Cross Examination

If the plaintiff's attorney engages
in redirect examination the defense
can elect to re-cross examination,
limiting the questions to topics
covered in redirect examination.

6.3.4.11 Objection

The opposing attorney can file a
number of motions or orally state
an objection while witnesses are

being questioned including an objection to a question, an answer, or both. If the judge considers the objection to be valid, the judge may support the motion and instruct the jury to ignore the statement or the question.

6.3.4.12 **Resting the Case**

Once the plaintiff presents evidence and all the witnesses have been fully questioned, the plaintiff's attorney will rest the case. At this point the defendant's attorney can decide whether to present the case for the defendant or to settle.

6.3.4.13 **Directed Verdict**

After the defendant decides to present his or her case, one or several strategic motions may be made, including a motion for a directed verdict that alleges that a defendant has failed to prove his or her case. If the directed verdict is granted, the plaintiff may not start another case on the same grounds.

6.3.4.14 **Voluntary Nonsuit**

If a plaintiff feels that a motion for a directed verdict will be granted by the judge, the plaintiff can react before resting the case by making a motion for a voluntary nonsuit. This would allow the plaintiff the right to begin another action on the same grounds after paying court costs.

6.3.4.15 **Case for the Defense**

If the plaintiff does not move for a voluntary nonsuit and a motion for a directed verdict is denied, the trial will proceed and the defense will present its case following the same procedures applicable to the plaintiff.

6.3.4.16 Rebuttal Evidence

After presenting the last witness the defendant may ask for a motion for a directed verdict, whether or not the defendant sought it before. If the motion is denied, the plaintiff may offer evidence to rebut what was said by the defendant's witnesses and subsequently the defendant may introduce evidence regarding that introduced during rebuttal.

6.3.4.17 Summation

After all the evidence has been heard and both sides rest, the plaintiff's attorney presents a summation whereby he or she recapitulates the plaintiff's claims, while commenting on the evidence. The defendant's attorney does likewise, and then the plaintiff's attorney is given the opportunity for rebuttal.

6.3.4.18 Charge to the Jury

After the attorneys have spoken, the judge charges the jury, pointing out important issues of law, and summarizing the testimony and how the jury should evaluate it. The judge may point out that the person bringing the claim has the burden of proof, with proof being based on the preponderance of evidence.

6.3.4.19 Requests to Charge

Either or both lawyers may present requests to charge, outlining special charges that they wish the judge to consider. Either attorney may also object to the charge as given.

6.3.4.20 The Verdict

After it is charged, the jury goes to the jury room for deliberation. If agreement cannot be reached, then a hung jury results and the case may be retried. However, when a verdict by the jury is reached it is presented to the court.

6.3.4.21 Motion for a New Trial

After the jury delivers its verdict, the attorney for the losing side may make a motion for judgment notwithstanding the NOV, a form of directed verdict. If it is not granted, the losing attorney can make a motion for a new trial. Such a motion could be granted for a number of reasons.

6.3.4.22 Final Judgment

If the judge's or jury's verdict is not set aside and no new trial is granted, the judge directs entry of a final judgment for the successful party.

6.3.5 STAGE IV: CIVIL LITIGATION POST-TRIAL PROCEEDING

A case is not closed after the final judgment is entered: Either party may enter an appeal in appellate court alleging errors were made at trial, an excessive award was given, or for other reasons.

6.3.5.1 Appellant

The appellant must notify the clerk of the court where the original trial was held that an appeal is being filed.

6.3.5.2 Appellee

The appellee must be notified by the appellant that an appeal is being filed.

6.3.5.3 Record of Appeal

The appellant's next step is to prepare a record of appeal. This identifies exactly why the appeal should be granted, citing whatever precedents may apply. The appellee also files a record of appeal, usually to prove the original verdict should stand.

6.3.5.4 Appellate Court

The appellate court consists of one or more judges, and the court's decision is based on their majority opinion. They affirm the lower court's opinion, reverse or modify the judgment, or grant a new trial. Once the judges reach their decision the case is sent back to the trial court for whatever action is necessary.

6.4 PRETRIAL RESPONSIBILITIES OF THE EXPERT WITNESS

The pretrial responsibilities of the forensic engineer include the investigation of the failure and preparation of the investigation report as described in previous chapters. Also included are a series of functions and responsibilities in preparation for the trial. These tasks directly involve the attorneys for the client.

6.4.1 PRETRIAL HEARINGS

The judge assigned to the case will usually assemble both sides at a pretrial hearing in order to simplify or streamline the case. The experts for both sides may be requested to attend by invitation or via subpoena. If the expert receives a subpoena, the client's attorney should be notified immediately if the expert's schedule conflicts with the hearing date.

6.4.2 DEVELOPING MATERIAL NEEDED FOR THE TRIAL: THE DISCOVERY PROCESS

When requested by the client's attorney, the expert should develop a list of documents

that are considered to be germane to the
trial preparation. The listings of documents
outlined in Section 2.5 "Document Search"
should be reviewed and a summary list
developed of all documents that will be
required for review and analysis of the
case. Coordinate this list of required
documents with the client's attorney to
ascertain if additional documents will be
required. In most cases the required
document may only be attainable through
discovery and a subpoena duces tecum.

The entire product of the expert witness in
preparation for the trial is usually protected by an extension of the client-
attorney privilege, at least until the
expert is officially designated as an
official expert witness.

The rule of privileged information varies
from jurisdiction to jurisdiction; the
client's attorney will clarify the situation. In most cases, the expert witness can
be requested to provide copies of all
material upon which his or her opinions and
conclusions are based at deposition, and
later at trial.

6.4.2.1 Interrogatories

When preparing a series of interrogatories to be posed to the
expert witness or some other party
on the opposing side, the attorney
for the client will usually request
the assistance of the forensic
expert in developing questions.
Also, the expert may be requested
to assist in responding to interrogatories posed by the other side,
and also may be requested to
review and analyze responses made
by the opposition to interrogatories. The responses to the
interrogatories made by the other
side may require additional
research or discovery activities on
the part of the expert after they
are reviewed.

6.4.2.2 Depositions

The expert will generally be deposed by the opposition's attorney. The process and technique for depositions is similar to that which occurs in court during civil litigation under cross-examination.

During the deposition the client's attorney will appear with the expert and advise on the proceedings. The attorney may have to present objections to certain questions or will advise the expert, off the record, of the potential pitfalls of answering certain questions.

A court reporter will be present at the deposition to develop a transcript of the proceedings. All witnesses at a deposition testify under oath. The transcript of the deposition will often be an important tool in preparing for trial and can be directly referenced during the progress of trial.

In preparing for a deposition, the expert should work closely with the attorney to develop a series of potential questions that may be asked by the other side, as well as a series of answers. Your answers should be honest, but answer only the questions asked. Do not volunteer information. If you do, you will probably provide more information than is requested and damage your client's case.

In preparing for the deposition of the opposing expert, it is generally requested that the expert assist the client's attorney in developing hypotheticals, questions that might be asked of the opposing expert or opposing plaintiff.

Request that you be advised by your client's attorney of the extent of deposition services that will be required. In addition to the deposition required by the opposing attorney, the forensic expert may be requested to attend the deposition of the opposing expert witness. Depositions can take extensive periods of time; the forensic expert must plan to allocate the time, which could amount to several weeks.

6.4.2.3 Preparation of trial exhibits

The use of models, graphics, sketches, and demonstrations in the courtroom will aid the forensic expert and the client's attorneys in explaining technical matters, and will assist in better presenting the opinions of the expert witness.

Visual presentations are important in presenting testimony because an individual's retention of visual information is nearly three times as much as retention of orally presented information. The use of visual aids will assist a jury of laypersons to comprehend the technical issues in a case.

If plans are to be exhibited it is better to prepare a diagram rather than using the originals. The use of color and other techniques, such as cross-hatching to differentiate one area from the other, is often desirable.

Graphics should be developed using color, whenever possible, and with large bold-face print. The graphics should be on a large scale and mounted on easels in a position that can be seen by a judge and jury. Provide reduced copies of the graphics to the judge and jury.

When developing a concept that can be best illustrated through a demonstration, be certain the

demonstration will work exactly as
planned before the trial starts.
Working models of concepts have
dramatic impact.

6.5   THE TRIAL: RESPONSIBILITIES OF THE EXPERT WITNESS

Activities of the expert witness during the trial
include preparing and rehearsing testimony;
preparing demonstrations, graphics and models;
assisting the client's attorneys in developing
direct and cross-examination testimony of others
on your side; as well as helping prepare the
cross-examination and hypotheticals the client's
counsel will use when questioning opposing
experts.

The ability of an expert witness to deliver
testimony in a clear and believable manner can be
enhanced by proper conduct when in the courtroom,
wearing proper attire, and answering questions in
a respectful and concise manner.

Following is a presentation of salient issues and
guidelines that are of importance to the expert
witness when preparing to deliver testimony during
the trial.

6.5.1 PLANNING

In planning for an appearance in court, a
number of issues should be identified and
resolved prior to the trial, including the
following:

6.5.1.1   Payment of Invoices

During cross-examination, the opposing attorney may ask questions
about the method of payment specified in your consulting agreement, or the attorney may ask if
your bill has been paid. Both
questions are intended to impeach
and discredit the expert's testimony. The expert's answer to the
former question should be that
payment is based on an hourly or
daily rate and you will receive
the professional fees for the service no matter what the outcome of
the case. Furthermore, it should

be stated that you are <u>not</u> working on a contingency basis. The latter question is asked so that in the event that your bill has not been paid in full, the opposing attorney can try to imply that your testimony may be shaded toward your client. Clear up your outstanding invoices prior to trial in order to avoid the opposing attorney's attempts at impeachment.

6.5.1.2 **On-Call Services**

The time period that must be set aside during a projected trial period, during which an expert witness may be requested to testify, is called on-call service. This blocking out of the expert's schedule could result in inconvenience for the expert by restricting travel plans, etc. When setting aside a large block of time while waiting to be called to testify, it may be appropriate to charge your client for the time expended while waiting. Charging a percentage of the regular hourly cost will defray some of the time lost that could have been charged working for other clients, and may result in better planning by the client's attorney.

6.5.1.3 **Visual Aids**

As previously discussed, visual aids are important in explicating the opinions and conclusions of the expert witness. The visual aids used to support the testimony of the expert witness must be prepared well in advance of the trial. The visual aids should be packed in proper carrying cases in order to protect them when travelling.

If slides or overhead projection graphics are to be used, review the physical arrangement of the courtroom in advance to assure it can accommodate the planned visuals; the courtroom must be capable of being darkened. Make provisions to assure that the sight lines for the judge and jury in the courtroom are clear, so that the slides or projections will be effective, and that the proper equipment is available. It is also advisable to carry an extra projection lamp.

6.5.1.4 Attire

The expert witness's attire and how it affects the opinion of the jury may vary somewhat depending on the geographical location. Speak with the client's attorney to ascertain what type of attire is recommended. In some jurisdictions, especially in rural or tropical settings, casual or sports clothing may be considered appropriate. The customary attire for the expert, however, is conservative dress: a dark suit, white shirt and shined shoes with a minimum of adornment. This means excluding flashy jewelry, fraternal pins, or other insignia.

The general rule is to dress in a fashion that denotes seriousness and professionalism to the lay-persons in the jury. It is important that the expert's attire create a dignified image of authority, and that the jury listen attentively to the expert's testimony rather than be distracted by the expert's dress.

6.5.1.5 Seating

The seating of the expert witness in the courtroom varies with different jurisdictions. In some courts, no witnesses, including experts, are allowed in the courtroom while opposing testimony is being presented. In some jurisdictions, the rule of exclusion does not apply to experts. In these cases, the client's attorney may prefer that the expert sit with them during the entire trial. Sitting next to the attorney permits the experts to comment on the testimony of opposing witnesses and experts.

Some attorneys do not require their experts to sit with them, but prefer for the expert witness to sit in the spectator section. These attorneys feel that when the expert sits next to them it appears that the expert is part of an advocacy team rather than an objective professional.

Consult the client's attorney to ascertain the rules of the juris-diction regarding the presence of expert witnesses in court. If applicable, determine if you are to sit with the client's attorney before and after giving expert testimony.

6.5.2 CONDUCT DURING TRIAL

The role of the forensic engineer as an expert witness during a trial that involves a judge and jury is not something that is experienced at college or graduate school, nor is it learned as a design professional. Thorough preparation and knowledge of the issues and a professional and relaxed demeanor are important factors in being a believable expert witness in court, in depositions, as well as other legal forums.

This section describes valuable techniques that can be used to improve performance when delivering testimony in court.

6.5.2.1 Attitudes and Demeanor

The personal behavior of the expert witness in the courtroom is important to the outcome of the trial. Presenting a positive attitude and proper demeanor during trial can mean the difference between appearing credible and presenting a pretentious and implausible image to the judge and the jury.

The client's attorney should assist the inexperienced expert in conveying a proper attitude and demeanor when in court or at a deposition. The following discussion can be of assistance to the novice as well as the veteran expert in creating a winning courtroom appearance.

6.5.2.1.1 Respect

Showing respect to the court and its proceedings will establish a positive appearance in the eyes of the judge. The expert witness should address the judge as "your honor" and respond to questions with "Yes sir" or "No sir."

6.5.2.1.2 Qualification

When the expert witness is asked questions by the client's attorney, it is important that the expert does not appear to be bashful or overly modest in reciting details about education, licenses, experience, professional society affiliations, and any other information that will aid in qualifying the witness as an expert. The forensic engineer has been retained as an expert and can only reduce credibility by being modest and downplaying accomplishments. On the other hand, if the expert is boastful and exaggerates his or her accomplishments, the expert should expect an attack by the opposing attorney. Boasting will most certainly reduce the credibility of the expert in the eyes of the judge and jury.

6.5.2.1.3   **Body Language**

Remain dispassionate and be constantly aware of body language. Maintaining a professional and positive attitude is critical to convincing the judge and jury of your sincerity. The judge and jury will react positively to proper body language when practiced by the expert witness.

The deportment of the expert witness and the proper and effective use of eye contact, facial expressions, and posture can create a positive rapport between the expert and the judge and jury.

6.5.2.1.4   **Posture**

The expert should sit up straight in the witness stand. Do not slouch or sit in a manner that may be construed as disrespectful.

6.5.2.1.5 **Gestures**

Avoid excessive or un-
necessary hand or head
gestures. Do not use the
hands to speak, es-
pecially when under
attack by the opposing
attorney during cross-
examination Do not nod
your head; all answers
must be made verbally
since the court reporter
cannot record a nod.

6.5.2.1.6 **Eye Contact**

The use of eye direction
and eye contact may vary
with the type of trial;
if it is a jury trial,
look directly at the
jury when answering
questions. If the ex-
pert can identify jurors
who seem uninterested,
it is recommended that
he or she make eye
contact with them. This
technique will assist
the expert in answering,
and will help him or her
develop a positive re-
lationship with the
jury. As a variation,
eye contact may be
shared with the judge
and jury in order to
establish an overall
rapport.

If there is not a jury,
then the expert should
maintain eye contact
with the judge when
answering.

6.5.2.1.7 **Facial Expression**

Do not use facial ex-
pressions to convey your
feelings; for instance,

do not scowl at the opposing attorney or react with facial expressions.

Most probably, the judge is or was a practicing attorney, and will not look favorably at the expert witness who is acting overtly negative toward the opposing attorney.

6.5.2.1.8 **Verbal Expression**

The technique of speaking clearly and pausing before answering are important to maintaining a professional attitude and presenting the image of an objective expert witness.

1. **Speaking Clearly**:
   The jury expects the expert, as an authority figure, to speak in a clear, concise and forceful voice. Delivering oral testimony in a clear, concise voice will create a positive image for the jury.

   Depending on the acoustics of the courtroom and whether or not a microphone is used, it may be necessary to use a "stage voice." A stage voice is used to project your normal voice at a level that can easily be heard by the judge, jury, and the court reporter.

It is important to clearly voice all your responses without gestures because the court reporter transcribes only what is stated during oral testimony. The outcome of the case may be determined based on the contents of the transcript.

2. **Pausing Before Answering**: When answering questions from either the client's attorney or the opposition's attorney, it is useful to pause momentarily before answering. The pause may be only a few seconds, but during it, you might want to look "thoughtful," in order to avoid a negative impact.

   Pausing is helpful for several reasons:

   1. Pausing before answering questions asked by the opposing attorney will afford the opportunity for your client's attorney to voice an objection to the question.

   2. Pausing before answering a question can be used to prepare an answer. (The witness might also say, "Would you

please rephrase
the question?")

3. Pausing before
answering during
stressful periods
of cross-examin-
ation will pre-
sent an oppor-
tunity to over-
come emotion.

6.5.2.1.9 <u>Maintaining Objectivity</u>

When testifying in court
it is paramount that the
expert witness maintain
an air of objectivity
and remain impartial and
dispassionate.

Following are guidelines
for techniques to assist
in maintaining objec-
tivity and impartiality
when testifying:

1. <u>**Maintain a dispassionate attitude.**</u>
When testifying dur-
ing trial, remain
dispassionate at all
times, no matter what
attempts the opposing
attorney may make to
insult your intelli-
gence.

The verbal attacks by
the opposing attorney
are usually intended
to excite the expert
and cause him or her
to become ruffled. If
the opposing attorney
is abusive, the
judge and jury will
see through these
tactics. The cli-
ent's attorney will
lodge objections to
the attacks. The

expert witness, by keeping his or her composure and remaining cordial to the opposing attorney, will appear professional, and the opposing attorney, using obnoxious tactics, will appear to be a bully.

2. **Avoid confrontation with opposing attorneys.** The expert is in court to render impartial testimony and to assist the judge and jury in understanding technical matters. The expert witness must avoid confrontations with opposing attorneys and avoid any impulse to try to outsmart the other side's attorney. Confrontations or engaging in a battle of wits will damage the expert's credibility.

6.5.2.1.11 Maintaining Perspective

The expert should maintain a sense of overall perspective in the case. Do not be overly concerned about creating misleading impressions in the testimony. It is possible that an expert will get the feeling that his or her testimony and the case are not going well. Do not adjust your testimony in an effort to correct what you perceive as wrong impressions. The client's attorney will tell you how things are going. The attorney will be like a football coach; if the attorney is not happy, then you will certainly know about it.

In addition, do not be concerned about how you have answered questions. If you feel that questions have been answered poorly, do not dwell on how to correct what you perceive to be misconceptions. You will lose your concentration and worsen the situation. Concentrate on each question asked. An opportunity to clarify misleading impressions will be offered during redirect testimony.

Remember, do not get rattled because you are dissatisfied with your past answers. Concentrate and give effective responses. Consult the client's attorney; the attorney may not agree that you've done poorly, and this can bolster your confidence in your testimony.

6.5.2.1.12 Behavior when not on the stand

The conduct of the expert witness in the courtroom when not giving testimony can be as important as conduct on the stand. Following are examples of well-considered behavior:

1. When sitting with the client's attorney, do not engage in extensive note passing unless the matter is urgent. It is advisable to make notes during the proceedings and discuss your thoughts with the attorney during a recess.

2. Pay strict attention to the proceedings and take notes. Do not read a newspaper or a novel, or do office work. To display obvious disrespect may prompt the judge to advise the expert to put away the reading matter. The judge may remember the discourtesy when the expert takes the stand; the jury certainly will.

3. Do not whisper in the court room.

4. If the opposing expert happens to be an acquaintance, do not appear to be overly friendly. The fact of your friendship may enhance the image of the opposing expert in the eyes of the judge and jury.

5 Attempt to maintain a stationary position in the courtroom. Do not enter and leave frequently.

6.5.3 <u>EFFECTIVE DELIVERY OF TESTIMONY</u>

Testimony given by an expert witness in a trial may be given either orally or in the form of prepared written testimony that can be read into the record. Technique for presenting effective testimony in court as well as guidelines for responding to questions when giving testimony are outlined in this section.

6.5.3.1 <u>Oral Testimony</u>

The most common form of testimony is given orally. Oral testimony is presented when the expert witness takes the stand, after being presented with a request for the expert to identify himself or herself. The expert then states his or her education, professional background and areas of special expertise.

The client's attorney will question the expert witness in a manner that is intended to develop a logical thought process that will serve to assist the judge and jury in understanding the technical issue involved.

The attorney is not permitted to ask leading questions of the expert but the attorney may pose theoretical questions that have been prepared in advance. These questions are intended to present the client's case in a manner that will familiarize the court with the opinions and the conclusions of the expert witness.

Testimony should be given by the expert witness in plain English without using technical jargon. The use of simple understandable terms while explaining technical issues, and the use of graphics and demonstrations will aid the court in understanding technical con-cepts.

In preparing for oral testimony, it is recommended that the expert witness read the transcripts of depositions of parties that are germane to the client's position. However, when on the stand, the expert should avoid reading directly from his or her own deposition. This could reduce the effectiveness of an expert since a layperson may feel that the expert is using the written text as a crutch.

The expert may prepare and use written notes while giving testimony. However, be aware that the opposing attorney or the judge can examine any notes that are brought to the stand. Be cautious of how written notes are prepared and expressed.

The opposing attorney will ask questions during cross-examination. These questions will generally be different than those posed during the deposition. If possible, work with the client's attorney to obtain advice on how the opposing attorney has conducted cross-examinations in similar trials. This is done to acquaint you with the opposing attorney's tactics and to avoid potential surprises arising from clever or disarming questions and other pressure techniques.

Cross-examination is more dramatic and could be more stressful than direct examination. The opposing attorney will attempt to discredit the expert during cross-examination. The expert witness must avoid the pitfall of turning cross-examination into a confrontation. The expert should keep "cool" and avoid falling into a battle of wits with the opposing attorney, because the attorney always has the last word.

If the client's attorney makes an
objection to questions during
cross-examination, do not answer
because the judge must either sus-
tain or overrule the objection. If
the objection is overruled, pause
before answering, and ask the opposing attorney to repeat the
question.

During cross-examination in a
court, effort may be made to discredit the expert witness. The
opposing attorney will try to
create an illusion of doubt about
the credibility of the expert witness. Knowing the traps that will
be laid down by the opposing attorney should help to keep the
expert from falling prey to them.
It is recommended that the inexperienced expert consult with the
client's attorney to identify these
tactics as well as how the opposing attorney may attempt to use
gratuitous impeachment to discredit
the expert witness by attempting to
establish contradiction in the
testimony of the expert.

6.5.3.2 **Written Testimony**

Written or prepared testimony can
be developed by an expert witness.
The final written report, which
serves as a chronicle of the investigation, can be inserted into
the record as evidence.

This document will be read by the
judge and a jury of laypersons. It
is important that this document be
written in clear language so it can
be understood by the jury.

6.5.4 **RESPONDING TO QUESTIONS DURING TESTIMONY**

The ability of an expert witness to deliver
testimony in a clear and honest manner
depends on experience as well as certain
techniques for answering questions. The
methodology for effectively answering ques-

tions differs depending on whether the
expert is under direct examination or
cross-examination.

These techniques are intended to assist your
client's attorney in developing the case and
to lend as little assistance to the opposing
attorney as possible.

6.5.4.1 Direct Examination

Voir Dire

The attorney will ask the court to
recognize the forensic engineer as
an expert witness. The attorney
will ask the forensic engineer to
state his or her credentials.
After the credentials have been
stated, the opposing attorney may
challenge the expert's credentials.

When the opposing attorney attacks
the credentials, it is imperative
that a calm composure be maintained
and all questions answered clearly. A challenge is seldom
successful, except for the purpose
of ruffling the expert witness.

6.5.4.1.2 Questions by the Client's Attorney

When responding directly
to the questions asked
by the client's attor-
ney, be complete in your
response, clearly state
in detail your opinions
and conclusions, and
carefully and concisely
present all background
data and analysis. Use
prepared visual aids if
necessary to demonstrate
findings.

6.5.4.1.3 Hypothetical Questions

When asked a hypothe-
tical question by the
client's attorney, give

a full recitation of the assumed facts and conclusions.

Hypothetical questions or "what if" statements are intended to clarify an important issue for the judge and jury, and will be developed and reviewed during preparation for trial.

6.5.4.1.4 **Objections**

Stop talking when an objection is issued by the opposing attorney, either to the client's attorney's questions or to your response. Do not resume speaking until the judge rules on the validity of the objection.

6.5.4.1.5 **Hearsay Evidence**

Avoid the use of hearsay evidence during testimony. Hearsay testimony is evidence not based upon the personal knowledge of the expert; it is evidence known to him or her only through another person's statements. Such evidence is generally inadmissible as testimony. Using hearsay testimony will bring objections from the opposing attorney. These objections could give the appearance that the expert's testimony is tainted.

6.5.4.1.6 Notes on Testimony

It may be helpful to use notes and data to assist in giving testimony but, remember, these notes may be requested for review by the opposing attorney. If so, the contents of the notes may be used against you by that attorney during cross-examination.

6.5.4.2 CROSS-EXAMINATION

6.5.4.2.1 Questioning by the Opposing Attorney

When being questioned by the opposing attorney, answer the questions briefly, without elaboration. Expanding on your answers may give the opposing attorney more information than is minimally required to answer the question and may aid the opposition's case.

Take your time when answering; do not rush your response. Think clearly before answering.

If you do not know the answer, state "I do not know." If you feel that your answer would be awkward, ask the opposing attorney to "Please rephrase that question." If you still do not understand, then state "I do not understand that question."

6.5.4.2.2 Objections by the Client's Attorney

If the client's attorney raises an objection to a question by the opposing attorney, do not answer until the court has ruled on the objection.

6.5.4.3 RE-DIRECT AND CROSS-EXAMINATION

6.5.4.3.1 Expanding or Clarifying Answers

In re-direct examination, your client's attorney may request that you expand or clarify certain responses that were made during cross-examination. This will give the expert the opportunity to clarify and expand responses that were given to questions posed by the opposing attorney. When expanding on your answers, respond in the same manner as described in direct cross-examination.

6.5.4.3.2 Answering During Re-Cross Examination

When being questioned during re-cross examination, conduct yourself the same way you did in cross-examination. Having just completed re-direct examination, where you cleared up all misconceptions about your earlier testimony, your responses can be made with renewed confidence.

6.6 POST-TRIAL RESPONSIBILITIES OF THE EXPERT WITNESS

The post-trial responsibilities of the forensic engineer include assistance to the client's attorney in preparing documents for appeals or for a new trial, as well as assistance in retaining investigation records and test specimens subsequent to trial for possible use in appeal.

CHAPTER 7

CONCLUSIONS

The need for a well organized, methodical approach to failure investigation arises from the complexity of modern engineered facilities and from the vast variety of possible failure causes. The investigative process should not employ a rigid "cookbook" approach, but there are certain steps that are common to an effective failure investigation. This guideline endeavors to provide the elements of a comprehensive failure investigation while not restricting the investigative process.

7.1 NEED FOR FAILURE INVESTIGATION

Failure of an engineered structure, assembly, component, etc., may range from a spectacular collapse to a relatively simple serviceability problem such as a drafty window. Failures are usually investigated in order to identify the responsible parties, to enable a proper repair and to avoid repeating the mistakes that led to the failure. Identification of the responsible parties is not to determine culpability, but rather to determine if the failure is related to such things as design, construction, and materials, etc.

In order for more engineers to learn from failures, a network to disseminate information is necessary. To accomplish this, the significant findings of a failure investigation must be condensed and made available to the profession at large. Currently there are several repositories for failure case studies. One data base that can be computer accessed is at the Architecture and Engineering Performance Information Center (AEPIC), developed by the University of Maryland School of Architecture and College of Engineering.

7.2 FUNDAMENTAL FAILURE INVESTIGATION PROCESS

While the types of failures and the elements/components that can fail are virtually unlimited in most engineered facilities, there are three fundamental components to the failure investigation process:

1. acquisition of data

2. analysis of data

3. presentation of opinions and conclusions

For minor failures, this process may be conducted by a single investigator over the course of a few hours. For a major collapse, however, this process may take teams of investigators representing various engineering disciplines several years.

A combination of on-site data collection, research, on and off-site testing, analysis, and documentation are common parts of a full scope failure investigation. Depending on the nature of the failure, some of these steps may be altered, expanded, or even omitted without compromising the investigation findings.

The investigation scope and budget are most often dictated by the economic consequences of the failure. Since failure hypotheses are developed and rejected based on data from costly field measurements/observations and tests, the degree of certainty obtained is often a function of the investigation budget. Even with a limited budget, however, the integrity of the investigation outcome will not be compromised provided that the budget and scope is adequate to provide a reasonable degree of certainty.

7.3   THE PRINCIPAL INVESTIGATOR

Forensic engineering is becoming an increasingly more specialized field, but it is not an engineering discipline in itself. The principal investigator should be best qualified in the discipline that is most closely related to the failure. In addition to a strong technical background, a broad experience in failure investigations is necessary for major investigations.

Reliance on past experience too early in the investigation when trying to determine the probable failure cause is a common critical error. Given the extreme range of possible failure causes, investigators who conclude before all of the hypotheses are explored that they have seen this type of failure before, often prejudices their judgment. Experience serves best when it aids the investigator in recognizing failure symptoms but not when it creates preconceptions that narrow the investigator's search.

Since the principal investigator tailors the specific failure investigation process based on the failure type or magnitude, the investigation outcome will only be as valid as the principal investigator's best judgment.

7.4 THE LEGAL ENVIRONMENT

Preparation for expert testimony is increasingly becoming a large part of the failure investigation. Now, in addition to technical competence, investigators must employ both methods and demeanor that are ever more persuasive. They must be able to convey technical meanings to nontechnical lawyers, judges, and arbitrators.

Often investigators are pressured by the client's attorneys to take an advocacy position. While this may cause the investigators to alter their reporting style, the substance of their opinions must not be compromised.

Many competent engineers avoid forensic engineering and failure investigation because of their dislike of the litigation process. The questioning of the investigation results and even of the investigator's competence may be completely distasteful to the engineer. A qualified engineer engaged in failure investigations must learn to cope with the rigors and requirements of the litigation process. This means fully documenting all significant data and thoroughly preparing for expert testimony early in the investigation.

7.5 OVERVIEW

An emphasis on an open-minded acceptance of all pertinent data while recognizing the relevant information is the fundamental basis of a failure investigation. As various hypotheses develop and change, continuous review of previously rejected data is required.

There is seldom a single cause of a failure but rather a complex interaction of components and forces. As a result, the outcome of a failure investigation seldom leads to absolutely irrefutable results but rather to a most probable cause of failure. While the collection of data and facts surrounding the failure are specific, expert opinions arising from these data may differ widely.

The most accepted failure investigation findings will be the one employing a qualified investigation team that presents the most plausible failure scenario based on well documented supporting data.

Acret, J., "Architects and Engineers: Their Professional Responsibilities", McGraw-Hill, New York, 1977.

American Society of Civil Engineers Committee on Forensic Engineering, "Report of the Working Group in Information Dissemination - Committee on Forensic Engineering - ASCE", March, 1984.

"Architects' and Engineers' Professional Liability Bibliography II". Guidelines for Improving Practice, Vol. 9, No. 2, 1979.

Bachner, John P., "Facing Down the Hired Gun", Journal of Performance of Constructed Facilities, ASCE, Vol. 2, No. 4, November, 1988.

Brown, J. Crozier, "Anatomies of Computer Disasters," Proceedings of the First International Conference on Computers in Civil Engineering. New York: American Society of Civil Engineers, 1981, p. 250.

Brown, Seymour W., P.E., "Types of Clients and Services", Consulting Engineers, (January 1984).

Carper, Kenneth L., ed., Forensic Engineering, Elsevier Science Publishers, New York, NY, 1988.

Carper, Kenneth L., ed., Forensic Engineering: Learning from Failures, ASCE, New York, NY, 1986.

Carper, Kenneth L., "Failure Information: Dissemination Strategies", Journal of Performance of Constructed Facilities, American Society of Civil Engineers, Vol. 1, No. 1, February, 1987.

Cohen, Stanley, "Consulting Engineering Practice Manual", American Consulting Engineers Council, 1982.

Collins, J. A., Failure of Materials in Mechanical Design: Analysis, Prediction, Prevention. New York: John Wiley & Sons, 1981.

"Disaster Preparedness", Report to Congress, Executive Office of the President, Office of Emergency Preparedness, Volume I, II, III, January, 1972.

Eaton, K. J., and C. J. Judge, "Tornado Damage to Buildings", June 26, 1973, Great Britain Dept. of Environment, Building Research Establishment, Current Paper 16/75, February, 1975, p. 3.

Elliott, Arthur L., "Falsework Failures: Can They Be Prevented?", Civil Engineering, American Society of Civil Engineers, Vol. 43, No. 10, October, 1973.

"Failure Patterns and Implications", Building Research Establishment Digest, No. 176, April, 1975, p. 3.

Fairweather, Virginia, "Bailey's Crossroads: A/E Liability Test", Civil Engineering, American Society of Civil Engineers, Vol. 43, No. 10, November, 1975.

Feld, Jacob, "Construction Failure", John Wiley & Sons, New York, 1968.

Feld, Jacob, "Reshoring of Multi-story Concrete Buildings", Concrete Construction, Vol. 19, No. 9, pp. 243-248, May, 1974.

Fitzsimons, Neal and Donald Vannoy, "Establishing Patterns of Building Failures", Civil Engineering, American Society of Civil Engineers, Vol. 54, No. 1, January, 1984.

Fitzsimons, Neal and Anatole Longinow, "Guidance for Load Tests of Buildings", Journal of the Structural Division, ASCE, July, 1975.

Flint, A. R., "Risks and their Control in Civil Engineering, "Proceedings of the Royal Society (London) A, 376, (1981), pp. 167-179.

Forsyth, Benjamin and Frank L. Stahl, "Throgs Neck Bridge: Why Did Its Deck Deteriorate?", Civil Engineering, American Society of Civil Engineers, Vol. 53, No. 7, July, 1983.

Florman, Samuel C., The Existential Pleasures of Engineering, New York: St. Martin's Press, 1976.

Florman, Samuel C., Blaming Technology: The Irrational Search for Scapegoats. New York: St. Martin's Press, 1981.

Garren, John H., "Steel Bridge Inspection Using Acoustic Crack Detector and Magnetic Crack Definer", Prepared for Federal Highway Admin., October, 1973, p. 34.

"Germany, Belgium and Los Angeles Have Mandatory Design Review for Major Structures", Civil Engineering, American Society of Civil Engineers, Vol. 48, No. 10, October, 1987.

Gjow, Odd E., "Durability of Reinforced Concrete Wharves Norwegian Harbors," Page Bros., Ltd., Norwich, England, 1968, p. 208.

Godfrey, Edward, "Engineering Failures and Their Lessons", Privately printed, 1924.

Godfrey, K. A., Jr., "Building Failures - Construction Related Problems and Solutions", Civil Enginerring, American Society of Civil Engineers, Vol. 54, No. 5, May, 1984.

Godfrey, E., "Engineering Failures and Their Lessons", Akron Superior Printing Co., 1924.

Gordon, J. E., The New Science of Strong Materials: Or Why You Don't Fall Through the Floor, Second Edition, Harmondsworth, Middlesex: Penguin Books, 1976.

Gordon, J. E., "Structures: Or Why Things Don't Fall Down", New York: Da Capo Press, 1981.

Griffiths, Hugh, "Report on the Inquiry into the Collapse of the Flats at Ronan Point, Canning Town, Great Britain", Ministry of Housing & Local Government, 1968, p. 71.

"Guidlines for the Professional Engineer as a Forensic Engineer", National Society of Professional Engineers, Washington, DC, 1972.

Haines, Daniel, W., "Forensic Engineering: What Role for ASCE?", Civil Engineering, American Society of Civil Engineers, Vol. 53, No. 7, July, 1983.

Hall, Peter, "Great Planning Disasters", Berkeley: University of California Press, 1982.

"High Alumnia Cement Concrete in Buildings", Structural Engineer, Vol. 10, No. 9, pp. 324-328, September, 1974.

"Inspection and Evaluation of Two Steels from the Silver Bridge", by Battele Memorial Institute, prepared for Department of Transportation, Federal Highway Admin., Bureau of Public Roads, 1970.

Jorgensen, J. L., and W. Larson, "Field Testing a Reinforced Concrete Highway Bridge", for North Dakota State Highway Dept., December, 1974, p. 46.

Keminetzku, Doy, "Structural Failures and How to Prevent Them", Civil Engineering, American Society of Civil Engineers, Vol. 46, No. 8, August, 1976.

Kemo, K. O., and J. G. A. Croll, "Role of Geometric Imperfections on Collapse of a Cooling Tower", Structural Engineering, Vol. 54, No. 9, January, 1976, pp. 33-37.

Khachaturian, Narbey, ed., Reducing Failures of Engineered Facilities, ASCE, New York, NY, 1985.

Leonard, D. R., "Dynamic Tests on Highway Bridges - Test Procedures and Equipment", Transport and Road Research Laboratory, Crawthorne, England, 1974, p. 32.

McKaig, Thomas B., Building Failures: Case Studies in Construction and Design, New York, McGraw-Hill, 1962, p. 261.

Mark, Robert, Experiments in Gothic Structure, The MIT Press, Cambridge, MA, 1982.

Mark, Robert, and William W. Clark, "Gothic Structural Experimentation", Scientific American, November, 1984.

Mark, Robert, "The Structural Analysis of Gothic Cathedrals", Scientific American, November, 1972.

Minor, Joseph E., and Lyon W. Beason, "Window Glass Failures in Windstorms", Paper 11834, Journal of the Structural Division, Proceedings of the American Society of Civil Engineers, Vol. 102, No. ST1, January, 1976.

Neill, C. R., et al., "Guide to Bridge Hydraulics", Project Committee on Bridge Hydraulics, Roads and Transportation Assoc. of Canada, University of Toronto Press, 1973, p. 191.

Norton, Harry N., "Handbook of Transducers for Electronic Measuring Systems", Prentice-Hall, Inc., Englewood Cliffs, NJ, 1969, p. 703.

O'Brien, J. J., "Construction Delay", Chaners Books International, Boston, 1976.

Orenstein, Glenn S., "Instant Expertise: A Danger of Small Computers," Civil Engineering, 54 (June 1984), pp. 50-51.

Pannell, J. P. M., "An Illustrated History of Civil Engineering", Thames and Hudson, London, 1964.

Petroski, Henry, "To Engineer is Human", St. Martin's Press, New York, 1985.

Petroski, Henry, "Reflections on a Slide Rule," Technology Review, 84 (February/ March 1981), pp. 34-35.

Petroski, Henry, "When Cracks Become Breakthroughs," Technology Review, 85 (August/September 1982), pp. 18-20.

Plewes, W. G., "Masonry Bibliography 1900-1977", PB285-987, National Technical Information Service, Springfield, VA, 1979.

Pritzker, Paul E., P.E., "Investigative Techniques", Consulting Engineer, (January, 1984).

"Professional Liability Loss Prevention Manual", Association of Soil and Foundation Engineers (1978).

Ropke, John C., "Concrete Problems: Causes and Cures", McGraw-Hill, New York, 1982.

Ross, Stephen S., "Construction Disasters: Design Failures Causes and Prevention", McGraw-Hill, 1984.

Schwartz, M., and N. F. Schwartz, "Engineering Evidence", McGraw-Hill, New York, 1982.

"Snell, L. M., "Proceedings", North American Masonry Conference, University of Colorado, Boulder, August, 1978, p. 73.

"Source Book in Failure Analysis", American Society for Metals, Metals Park, Ohio, 1974, p. 406.

"Standard Forms of Agreement", Engineers' Joint Contract Document Committee.

Stanley, C. Maxwell, "The Consulting Engineer", John Wiley and Sons, 1982.

Stockbridge, Jerry G., "Cladding Failures - Lack of a Professional Interface", Paper 150-85, Journal of the Technical Councils, Proceeding of the American society of Civil Engineers, Vol. 105, No. TC2, December, 1979.

"Structural Failures, Modes, Causes, and Responsibilities", American Society of Civil Engineers, New York, 1973.

"Symposium on Concrete Construction in Aqueous Environments", Pub. SP-8, American Concrete Institute, Detroit, Michigan, 1964.

"Symposium Tags Ignorance as Root of Facade Problems", Engineering News Record, 11 December 1980, p. 17.

U.S. House of Representatives Committee on Science and Technology, Structural Failures: Hearings Before the Subcommittee on Investigations and Oversight. Washington, DC: Government Printing Office, 1983.

Vannoy, Donald W., Report to Executive Director, American Society of Civil Engineers, from Architecture and Engineering Performance Information Center, November 12, 1984.

Vannoy, Donald W., "20/20 Hindsight with the Aid of a Computer", Professional Engineer, 53 (Winter 1983), pp. 21-25.

211

LEGAL REFERENCES

Aidlin, S. S., P.E., "The Engineer as an Expert Witness", Professional Engineering, Economics and Practice (1952).

Aidlin, S. S., P.E., "The Expert Witness Prepares for Court," Steven Howard Company.

American Bar Association Directory, ABA, 1155 East 60th Street, Chicago, IL 60637.

Brown, Seymour W., P.E., Consulting Engineer, "Types of Clients and Services,", (January 1984).

Cantor, Benjamin J., "The Expert Witness", American Bar Association Journal, (October 1966).

Colby, E. E., "Practical Legal Advice for Builders and Contractors", Prentice-Hall, Englewood Cliffs, NJ, 1972.

Construction Failures: Legal and Engineering Perspectives, ABA/ASCE, Symposium in Houston, Texas, (October 1983).

The Consulting Engineer as an Expert Witness, Consulting Engineer's Association of California, (1983).

Cushman, Robert F., "Avoiding Liability in Architecture, Design and Construction", John Wiley & Sons, (1983).

"Design Professional Liability Claims", Guidelines for Improving Practice, Vol. 7, No. 8, 1977.

Dunham, C. W., "Contracts, Specifications and Law for Engineers", McGraw-Hill, New York, 1971.

Fairweather, Virginia, "Bailey's Crossroads: A/E Liability Test", Civil Engineering, American Society of Civil Engineers, Vol. 45, No. 11, November, 1975.

Goodkind, Donald R., "Mediation of Construction Disputes", Journal of Performance of Constructed Facilities, American Society of Civil Engineers, Vol. 2, No. 1, February, 1988.

Hanley, Robert F., "Working the Witness Puzzle", Journal of the Section of Litigation, American Bar Association.

Henry, James F., "ADR and Construction Dispute", Journal of Performance of Constructed Facilities, American Society of Civil Engineers, Vol. 2, No. 1, February, 1988.

Hohns, H. M., "Preventing and Solving Construction Contract Disputes", Van Nostrand Reinhold, New York, 1979.

Huston, John, P.E., "Engineers on the Witness Stand: Guidelines for Expert Testimony", Civil Engineering, (February 1979).

Knight, Warren H., "Use of Private Judges in Alternate Dispute Resolution", Journal of Performance of Constructed Facilities, American Society of Civil Engineers, Vol. 2, No. 1, February, 1988.

Kornblum, Guy O., "The Expert as Witness and Consultant", The Practical Lawyer, (March 1974).

Kraft, M. D., "Using Experts in Civil Cases", Practicing Law Institute, New York, 1980.

Legal Briefs for Architects, Engineers and Contractors, McGraw-Hill, New York, 1980.

Leval, Pierre N., "Discovery of Experts Under the Federal Rules".

McElhaney, James W., "Trial Notebook, An Outline on Hearsay", Journal of the Section of Litigation, American Bar Association, (Winter, 1978).

McElhaney, James W., "Trial Notebook, Foundations", Journal of the Section of Litigation, American Bar Associates (Spring, 1978).

McHenry, Douglas F. and Reese, Raymond C., "The Case for Evaluating Performance of Structures", Civil Engineering, American Society of Civil Engineers, Vol. 38, No. 3, March, 1968.

McQuillan, Joseph A., Esq., "The C.E. as an Expert Witness", Consulting Engineer, (January 1984).

Marcus, Stephen D., "Goals and Objectives for Alternate Dispute Resolution", Journal of Performance of Constructed Facilities, American Society of Civil Engineers, Vol. 2, No. 1, February, 1988.

Mazur, Sayward and Robert A. Rubin, "Construction Cases - Using Experts in Civil Cases", Practicing Law Institute, (1977).

Mazur, Sayward, Robert A. Rubin, Robert N. Shiverts, and Stanley J. Siegelbaum, "The Engineer as an Expert Witness", The (NYC) Municipal Engineer's Journal, (1977).

Meehan, Richard S., "Getting Sued (and Other Tales of the Engineering Life)", the MIT Press, (1981).

Nelson, Sandra L., "ADR - A Different Ritual: An Insurer's Perspective", Journal of Performance of Constructed Facilities, ASCE, Vol. 1, No. 4, November, 1987.

On Being an Expert Witness, Nelson Chalmers, Inc., 725 Teaneck Road, Teaneck, NJ, (August 1983).

Panalon, Eugene I., "ADR Trial Lawyers Perspective", Journal of Performance of Constructed Facilities, ASCE, Vol. 1, No. 4, November, 1987.

213

Pagan, Alfred R., "Ten Commandments (More or Less) for the Expert Witness", Better Roads Magazine, (1982).

Poirot, James W., "Alternate Dispute Resolution Techniques, Design Professionals Perspective", Journal of Performance of Constructed Facilities, ASCE, Vol. 1, No. 4, November, 1987.

Pritzer, Paul E., P.E., "Investigative Techniques", Consulting Engineer, (January 1984).

Professional Liability Loss Prevention Manual, Association of Soil and Foundation Engineers, 1978.

Qualification of Expert on Concrete/Proof that Damage to Structures Caused by Settling of Concrete/Proof of Faulty Placement of Concrete Reinforcement.

Richards, Frederick S., "Criminal Liability of the Field Engineer", Paper 13265, Engineering Issues - Journal of Professional Activities, Proceedings of the American Society of Civil Engineers, Vol. 103, No. E14, October, 1977.

Ritcher, Irv and Roy S. Mitchell, "Handbook of Construction Law and Claims", Reston Publishing Co., Reston, VA, (1982).

Rossi, Faust, "Evidence Highlights", NITA, St. Paul, MN.

Schwartz, M. and Schwartz, N. F., "Engineering Evidence", McGraw-Hill, New York, 1982.

Schweitzer, Sydney C., "So You're Going to Be a Witness?", Case and Comment.

Siegelbaum, Stanley J., "Some Aspects of the Law of Evidence".

Spector, Marvin M., P.E., "Engineering Applied to Jurisprudence", Consulting Engineer, (January 1984).

"Standard Forms of Agreement", Engineers' Joint Contract Document Committee.

Sweet, J., "Legal Aspects of Architecture, Engineering and the Construction Process", West Publishing Company, St. Paul, MN, 1970.

Thomason, B. and Coplan, N., "Architectural and Engineering Law", Reinhold Publishing Company, New York, 1967.

Vannoy, Donald W., "20/20 Hindsight, Overview of Failures", Proceedings of the Conference, "Construction Failures: Legal and Engineering Perspectives", American Bar Association, October, 1983.

Walker, N. and T. K. Rohdenbure, "Legal Pitfall in Architecture, Engineering and Building Construction", McGraw-Hill, New York, 1978.

Weinstein, A. S., et al, "Product Liability: A Study of the Interaction of Law and Technology", National Science Foundation, Washington, DC, 1977.

STRUCTURAL REFERENCES

American Concrete Institute, "Seminar Course Manual: Lessons from Failures of Concrete Buildings", American Concrete Institute, Undated.

American Society of Civil Engineers Research Council on Performance of Structures, "Structural Failures: Modes, Causes, Responsibilities", American Society of Civil Engineers, New York, 1973.

Anon, "Collapse of U.S. 35 Highway Bridge, Point Pleasant, West Virginia, December 14, 1967", National Transportation Safety Board, Report No. NTSB-HAR-71-1.

"Automobile Collision with and Collapse of the Yadkin River Bridge Near Siloam, North Carolina, February 23, 1975, National Transportation Safety Board, Report No. NTSB-HAR-76-3, April 1976, p. 30.

Babb, A. O. and T. W. Mermel, "Catalog of Dam Disasters, Failures, and Accidents", Bureau of Reclamation, 1968, p. 216.

Bares, Richard and Neal FitzSimons, "Load Tests of Building Structures", Journal of the Structural Division, ASCE, May 1975.

"Behind the Sines, Portugal Breakwater Failure", Civil Engineering, American Society of Civil Engineers, Vol. 52, No. 4, April, 1982.

Bell, Glen R. and Parker, James C., "Roof Collapse, Magic Mart Store, Bolivar Tennessee", Journal of Performance of Constructed Facilities, ASCE, Vol. 1, No. 2, May, 1987.

Blockley, D. I., "The Nature of Structural Design and Safety" Chichester, West Sussex: Ellis Horwood Limited, 1980.

Bowers, D. G., "Loading History Span No. 10, Yellow Mill Pond Bridge I-95, Bridgeport, Connecticut", State of Connecticut Department of Transportation, Research Report No. HPR 175-332, Weathersfield, Connecticut, May, 1972.

Boyd, G. M., "Brittle Fracture in Steel Structures", Butterworth & Co., Ltd., London, 1970, p. 122.

Brow, B. F., "Stress-Corrosion Cracking and Related Phenomena in High Strength Steels: A Review of the Problem with an Annotated Bibliography", U.S. Navy Research Lab, NRL Dept. 6041, November, 1953, p. 1963, p. 22.

Bruinette, Konsant E., "Kazerne Viaduct Collapse," Paper 10361, Journal of the Structural Division, Proceedings of the American Society of Civil Engineers, Vol. 100, No. ST2, February, 1974.

Byrne, Robert, "Skyscraper", New York: Atheneum, 1984.

Carino, Nicholas J., H. S. Lew, and William C. Stone, "Investigation of East Chicago Ramp Collapse", Journal of Construction Engineering and Management, Proceedings of the American Society of Civil Engineers, Vol. 110, No. 1, March 1984.

Carper, Kenneth L., "Structural Failures During Construction", Journal of Performance of Constructed Facilities, ASCE, Vol. 1, No. 4, November, 1987.

Carter, S. C., M. V. Hyarr, and J. E. Cotton, "Stress-Corrosion Susceptibility of Highway Bridge Construction Steels", prepared for the Federal Highway Administration, 1972, p. 300.

Champion, S., "Failure and Repair of Concrete Structures", New York, John Wiley and Sons, 1961, p. 199.

Choate, Pat and Walter, Susan, "America in Ruins: The Decaying Infrastructure", Durham, NC: Duke University Press, 1983.

Clear, Ken, "Permanent Bridge Deck Repair", Public Road, Vol. 39, No. 2, September 1975, pp. 53-62.

"Collapse of U.S. 35 Highway Bridge, Point Pleasant, West Virginia, December 14, 1967", National Transportation Safety Board, Report No. NSTB-Har.-71-1.

"Construction Failures: Legal and Engineering Perspectives", ABA/ASCE Symposium in Houston, Texas (October 1983).

"Construction of Metals on Concrete", Publication SP-49, p. 136, American Concrete Institute, Detroit, Michigan, 1975.

Creasy, Leonard R., "Prestressed Concrete Cylindrical Tanks", John Wiley & Sons, Inc., New York, 1961, p. 216.

Culver, Charles G. and Scribner, Charles F., "Investigation of the Collapse of L'Ambiance Plaza", Journal of Performance of Constructed Structures, ASCE, Vol. 2, No. 2, May, 1988.

Czyzewski, Harry and Earl C. Sutherland, "The Use and Misuse of Metal Structures", Civil Engineering, American Society of Civil Engineers, Vol. 38, No. 8, August, 1968.

Dicker, Daniel F., "Point Pleasant Bridge Collapse Mechanism Analyzed", Civil Engineering, American Society of Civil Engineers, Vol. 41, No. 7, July, 1977.

Dickey, Walter L. and Glen B. Woodruff, "Vibrations of Steel Stacks", Transact of American Society of Civil Engineers, Vol. 121, pp. 1054-1070, 1956.

"Differential Movement: Cause and Effect", Technical Notes on Brick and Tile Construction, No. 18, Structural Clay Products Institute, April, 1963, p. 16.

Downey, Eric and Tain A. MacLead, "Serviceability Failures Due to Structural Movements", Journal fo the Technical Councils, American Society of Civil Engineers, Vol. 108, No. TCI, May, 1982.

Eaton, Keith J., "Cladding and the Wind", Paper 12114, Journal of the Structural Division, American Society of Civil Engineers, Vol. 102, No. ST5, May 1976.

Farguharson, F. B., "Aerodynamic Stability of Suspension Bridges, Model Investigation Which Influenced the Design of the New Tacoma Narrows Bridge, Part IV", University of Washington Engineering Experiment Station, Bulletin No. 116, Part IV, 1954, p. 103.

Fattal, S. G. and L. E. Cattaneo, "Evaluation of Structural Properties of Masonry in Existing Buildings", BSC 62, U.S. National Bureau of Standards, Washington, DC, 1977.

Feld, J., "Lessons from Failures of Concrete Structures", American Society of Civil Engineers, Vol. 43, No. 6, June, 1973.

"Final Report on the Investigation Committee of the Malpasset Dam", Vol. I, Paris 1960. Translated from French, Published for the U.S. Department of Interior and National Science Foundation, Washington, DC by the Israel Program for Scientific Translations, Printed in Jerusalem by S. Monson. Availability from Office of Technical Service, U.S. Dept. of Commerce, Washington, D.C.

Findlay, W. P., "Dry Rot and Other Timber Troubles", Hutchinson's Scientific and Technical Publications, London, 1953, p. 267.

FitzSimons, Neal, "Structural Failure", Paper 15625, Journal of the Technical Councils, Proceedings of the American Society of Civil Engineers, August, 1980.

Gatge, Scott P., "The Role of the Federal Government in the Investigation of Structural Failures", WISE Internship Report to the American Society of Civil Engineers, August 3, 1984.

Godfrey, Kneeland, A., Jr., "Structural Failures", Civil Engineering, American Society of Civil Engineers, Vol. 51, No. 12, December, 1981.

Gurfinkel, German, "Tall Steel Tanks: Failure, Design and Repair", Journal of Performance of Constructed Facilities, ASCE, Vol. 2, No. 2, May, 1988.

Gurfinkel, German, "Precast Concrete Roof Structure: Failure and Repair", Journal of Performance of Constructed Facilities, ASCE, Vol. 2, No. 3, August, 1988.

Hahn, Oscar M., "Stability Problems of Wood Truss Bridge", Paper 7095, Journal of the Structural Division, Proceedings of the American Society of Civil Engineers, Vol. 96, No. ST2, February, 1970.

Hauck, George F. W., "Hyatt Regency Walkway Collapse: Design Alternatives", Journal of Structural Engineering, Proceedings of the American Society of Civil Engineers, Vol. 109, No. 5, May, 1983.

Hertzberg, R. W., "Deformation and Fracture Mechanics of Engineering Materials", New York: John Wiley & Sons, 1976.

Hetenyi, M., "Handbook of Experimental Stress Analysis", John Wiley and Sons, Inc., New York, NY, 1950, p. 1077.

Janney, Jack R., "Guide to Investigation of Structural Failures", New York: American Society of Civil Engineers, 1979.

Koster, Waldemar, "Expansion Joints in Bridges and Concrete Roads", Transatlantic Arts, Inc., New York, 1969, p. 333.

Kocsis, Peter, "Some Proposals for Reducing Structural Failures", Civil Engineering, American Society of Civil Engineers, Vol. 52, No. 6, June, 1982.

Leone, A., C. McGogney, and J. Barton, "A New System for Inspecting Steel Bridges for Fatigue Cracks", Public Roads, Vol. 37.

LePatner, Barry B., and Sidney M. Johnson, Structural and Foundation Failures: A Casebook for Architects, Engineers and Lawyers, McGraw-Hill, New York, 1982.

Lew, H. S., N. J. Carino, and S. G. Fattal, "Cause of the Condominium Collapse in Cocoa Beach, Florida", Concrete International, American Concrete Institute, August, 1982.

Lew, H. S., "West Virginia Cooling Tower Collapse Caused by Premature Form Removal", Civil Engineering, American Society of Civil Engineers, Vol. 50, No. 2, February, 1980.

Longbottom, K. W. and G. P. Mallet, "Prestressing Steels", Structural Engineer, Vol. 51, No. 12, pp. 455-471, December, 1973.

Loomis, Robert S., Raymond H. Loomis, Robert W. Loomis, and Richard W. Loomis, "Torsional Buckling Study of Hartford Coliseum", Paper 15124, Journal of the Structural Division, Proceedings of the American Society of Civil Engineers, Vol. 106, No. STI, January, 1980.

McHenry, Douglas F. and Raymond C. Reese, "The Case for Evaluating Performance of Structures", Civil Engineers, American Society of Civil Engineers, Vol. 38, No. 3, March, 1968.

Mason, A. Hughlett and Dean S. Carder, "Vibration Frequencies of the Chesapeake Bay Bridge", Journal of the Structural Division Proceedings of the American Society of Civil Engineers, Vol. 93, No. ST2, April, 1967, pp. 237-245.

Marshall, R. D., et al, "Investigation of the Kansas City Hyatt Regency Walkways Collapse", (NBS Building Science Series 143), Washington, DC: U.S. Dept. of Commerce, National Bureau of Standards, 1982.

Mather, B., "Factors Affecting Durability of Concrete in Coastal Structures", Technical Memorandum No. 96, Beach Erosion Board, Washington, D.C.

Monefore, G. E. and G. J. Verbeck, "Corrosion of Prestressed Wire", Journal of the American Concrete Institute, Vol. 32, No. 5, pp. 491-515, November, 1960.

Morton, J. and C. Ruiz, "Floating Roof Tank Design Is Eased", Gas and Oil Journal, April 12, 1976, pp. 59-61.

Paris, P. C. and G. C. M. Sih, "Stress Analysis of Cracks", ASTM, STP 381, 30, 1965.

Petroski, Henry, "On 19th Century Perceptions of Iron Bridge Failures", Technology and Culture, 24, (1983), pp. 655-659.

218

Pfrang, Edward O. and Richard Marshall, "Collapse of the Kansas City Hyatt Regency Walkways", Civil Engineering, American Society of Civil Engineers, Vol. 52, No. 7, July, 1982.

Pugsley, Sir Alfred, "The Safety of Structures", London: Edward Arnold, 1966.

Rahman, Mohammed A., "Damage to Karnafuil Dam Spillway", Paper 9452, Journal of the Hydraulics Division, Proceedings of the American Society of Civil Engineers, Vol. 98, No. HY12, December, 1972.

Reese, Raymond C., "Structural Failures from the Human Side", Civil Engineering, American Society of Civil Engineers, Vol. 43, No. 1, January, 1973.

Rhodes, Peter S., "The Structural Assessment of Building Subjected to Bomb Damage", Structural Engineer, Vol. 52, No. 9, pp. 329-339, September, 1974.

Roberts, Richard and George Irwin, "Fatigue and Fracture Mechanics of Bridge Steels", Journal of the Structural Division, Proceedings of the ASCE, Vol. 102, No. ST2, pp. 337-353.

Ross, Steven S., "Construction Disasters: Design Failures, Causes and Prevention", New York: McGraw-Hill Book Company, 1984.

Silby, P. G., and A. C. Walker, "Structural Accidents and Their Causes", Proceedings of the Institution of Civil Engineers, Part 1, 62 (May 1977), pp. 191-208.

Smith, Denis, "the Use of Models in Nineteenth Century British Suspension Bridge Design", History of Technology: Second Annual Volume, 1977, pp. 169-214.

Tedesko, Anton, "How Have Concrete Shell Structures Performed? An Engineer Looks Back at Years of Experience with Shells," Bulletin of the Internationa Association for Shell and Spatial Structures, 21 (August 1980), pp. 3-13.

Vose, George L., "Bridge Disasters in America: The Cause and the Remedy", Boston: Lee and Shepard, 1887.

Whyte, R. R., ed., "Engineering Progress through Trouble", London: The Instituteion of Mechanical Engineers, 1975.

Zallen, Rubin M., "Roof Collapse Under Snow Drift Loading and Snow Drift Design Criteria", Journal of Performance of Constructed Facilities, ASCE, Vol. 2, No. 2, May, 1988.

Zetlin, Lev Associates, "Report of the Engineering Investigation Concerning the Causes of the Collapse of the Hartford Coliseum Space Truss Roof on January 18, 1978", New York: Lev Zetlin Associates, 1978.

GEOTECHNICAL REFERENCES

Bell, Roy A. and Jun Iwakiri, "Settlement Comparison Used in Tank-Failure Study", Paper 15219, Journal of Geotechnical Engineering Division, Proceeding of the American Society of Civil Engineers, Vol. 106, No. GT2, February, 1980.

Daniel, David E. and Roy Olson, "Failure of an Anchored Bulkhead", Journal of the Geotechnical Engineering Division, Proceedings of the American Society of Civil Engineers, Vol. 108, No. GT10, October, 1982.

Duncan, James M. and Albert L. Buchigani, "Failure of Underwater Slope in San Francisco Bay", Paper 10019, Journal of the Soil Mechanics and Foundations Division, Proceedings of the American Society of Civil Engineers, Vol. 99, No. SM9, September, 1973.

Duncan, James M. and William N. Houston, "Estimating Failure Probabilities for California Levees", Journal of Geotechnical Engineering, ASCE, Vol. 109, No. 2, February, 1983.

Feld, Jacob F., "Foundation Failure", Civil Engineering, American Society of Civil Engineers, Vol. 43, No. 6, June, 1973.

Gnaedinger, John P., "Open Hearth Slag - A Problem Waiting to Happen", Journal of Performance of Constructed Facilities, ASCE, Vol. 1, No. 2, May, 1987.

Gnaedinger, John P., "Case Histories - Learning From Our Mistakes", Journal of Performance of Constructed Facilities, ASCE, Vol. 1, No. 1, February, 1987.

Ingold, Terrence S., "Retaining Wall Performance During Backfilling", Paper 14580, Journal of the Geotechnical Engineering Division, Proceedings of the American Society of Civil Engineers, Vol. 106, No. GT5, May, 1979.

Jeyapalan, Jey K., Michael J. Duncan, and H. Bolten Seed, "Investigation of Flow Failures of Tailings Dams", Journal of Geotechnical Engineering, American Society of Civil Engineering, Vol. 109, No. 2, February, 1983.

Khan, Idbal H., "Failure of an Earth Dam: A Case Study", Journal of Teotechnical Engineering, American Society of Civil Engineers, Vol. 109, No. 2, February, 1983.

Lambe, William T., Francisco Silva, and Allen W. Marr, "Instability of Amuay Cliffside", Paper 16636, Journal of the Geotechnical Engineering Division, Proceedings of the American Society of Civil Engineers, Vol. 107, No. GT11, November, 1981.

Leonards, Gerald A., "Investigation of Failures", Journal of the Geotechnical Enginering Division, Proceedings of the American Society of Civil Engineers, Vol. No. GT2, February, 1982.

LePatner, Barry B. and Sidney M. Johnson, Structural and Foundation Failures: A Casebook for Architects, Engineers, and Lawyers, McGraw-Hill, New York, 1982.

Nordlund, Raymond L. and Don U. Deere, "Collapse of Fargo Grain Elevator", Paper 7172, Journal of the Soils Mechanics and Foundations Division, Proceedings of the American Society of Civil Engineers, Vol. 96, March, 1970.

Peguignot, C. A., "Tunnels and Tunneling", Hutchinson & Co., Ltd., London, 1963, p. 555.

Sealy, Curtis O. and Bandimere, Sampson W., "Grouting in Difficult Soil and Weather Conditions", Journal of Performance of Constructed Facilities, ASCE, Vol. 1, No. 2, May, 1987.

"Scour at Bridge Waterways", Highway Research Board, 1970, U.S. National Cooperative Highway Research Program, Synthesis of Highway Practice, No. 5.

Szecny, Karoly, "The Art of Tunneling", Akademial Kiodo, Budapest, Hungary, 1967.

Szechy, C., "Foundation Failure", Concrete Publications, Ltd., London, 1961, p. 115.

Terzaghi, Karl and Ralph B. Peck, "Soil Mechanics in Engineering Practice", John Wiley and Sons, Inc., 1948, p. 566.

Terzaghi, Karl, "Failure of Bridge Piers Due to Scour", Proceedings of International Conference on Soil Mechanics, Vol. II, p. 264, Cambridge, MA, 1936.

Thompson, Louis J. and Ronald J. Tanenbaum, "Survey of Construction Related Tranch Cave-Ins", Journal of the Construction Division, Proceedings of the American Society of Civil Engineers, Vol. 103, CO3, September, 1977.

Tschebotarioff, Gregory, P., "Soil Mechanics, Foundations and Earth Structures", New York, McGraw-Hill, 1951, p. 655.